BEEKEEPING

The Beginning Beekeepers Guide to Their First Hive

(Comprehensive Guide for Indoor and Outdoor Organic Gardening and Beekeeping)

Ruby Hicks

Published by Harry Barnes

Ruby Hicks

All Rights Reserved

Beekeeping: The Beginning Beekeepers Guide to Their First Hive (Comprehensive Guide for Indoor and Outdoor Organic Gardening and Beekeeping)

ISBN 978-1-77485-184-5

Legal & Disclaimer

The information contained in this book is not designed to replace or take the place of any form of medicine or professional medical advice. The information in this book has been provided for educational and entertainment purposes only.

The information contained in this book has been compiled from sources deemed reliable, and it is accurate to the best of the Author's knowledge; however, the Author cannot guarantee its accuracy and validity and cannot be held liable for any errors or omissions. Changes are periodically made to this book. You must consult your doctor or get professional medical advice before using any of the suggested remedies, techniques, or information in this book.

Upon using the information contained in this book, you agree to hold harmless the Author from and against any damages, costs, and expenses, including any legal fees potentially resulting from the application of any of the information provided by this guide. This disclaimer applies to any damages or injury caused by the use and application, whether directly or indirectly, of any advice or information presented, whether for breach of contract, tort, negligence, personal injury, criminal intent, or under any other cause of action.

You agree to accept all risks of using the information presented inside this book. You need to consult a professional medical practitioner in order to ensure you are both able and healthy enough to participate in this program.

Table of Contents

Introduction

Apiculture is the practice of keeping bees. It involves the cultivation of colonies. It has a rich and varied history that dates back thousands of years. Beekeeping is an industry that has not changed in many ways since ancient times. However, it is still a vital industry and a lucrative hobby.

When beekeeping is done commercially, it involves the production and sale beeswax and honey, as well as the breeding of bees for later sale. For crop pollination, beekeepers can even rent bees.

Bee products sold sometimes include royal jelly, bee pollen and bee venom.

Apiculture can also be done as a hobby. The beekeeper might decide to take part in any or all the activities listed above.

People began to cultivate crops and live in villages. Beekeeping became an important way to provide sweetener for the village. Honey was the only sweetener that man knew of, until sugarcane was introduced to sweeten it.

As early as 2,000 BC, the first cultivated hives were found in Egypt and China. Honey hunters used to hunt wild bee nests in order to collect honey and keep hives. Scientists discovered cave drawings that show honey collection from rocks and trees as far back as 7,000 BC.

Some places around the globe, beekeeping has not changed that much. As in ancient times, bees can still be found in hollow logs and wicker baskets, barrels, gourds, and even in cavities within walls. Honey has been used as a sweetener and as an ingredient in alcoholic beverages mead.

The importance of hives increased with the spread of Christianity. For the production of candles, beeswax was highly in demand. Today, beeswax continues to be used in the cosmetics and healthcare industries.

The Greek method of beekeeping is the basis for modern beekeeping. The Greeks invented a "bee area" and used wicker baskets to house their hives. The hive's opening was covered with wood bars at distances of 6.350 to 9.525mm (or.25-.375 inches).

The bees couldn't build combs between the bars because of the distance between them. This allowed beekeepers the freedom to remove combs from hives to inspect them and/or to create new hives using the combs.

George Wheler, in 1682, described this method of building hives. This description formed the basis for the design of hives in England, and Europe thereafter.

Today's hives have a board at the bottom, a collection of boxes with movable frames, and a cover. Each frame has a beeswax bottom or a plastic bottom imprinted with hexagonal cell bottom shapes. The shapes are used by the bees as guides, and they build their combs according the pattern of the cell tops.

This beginners guide will help you learn the basics of beekeeping.

Chapter 1: Beekeeping Basics

What does a beekeeper do? An apiarist is a person who manages bee colonies to harvest honey and other products. A beekeeper will collect honey from bees that have more than they need to survive. This honey is the most nutritious and local sweetener. About 30% of honey consumed in America is made from this honey. Bee colonies, which are most often found in hives produce pollen, pollen and propolis. The term "apiary", or "bee yard", refers to the area where bees are kept.

Honey can be used in many ways. In a later chapter, we will discuss the many uses of honey and its health benefits. You can use beeswax for cosmetics, wood polish, candles and even wood polish. You can also use pollen to pollinate crops. Propolis, also known as bee glue, can be used to make car wax, varnish for musical instruments and as an ingredient in gum. Royal jelly has many nutritional and medicinal uses.

Anyone can be a beekeeper. It is easier to be a beekeeper if you live in a rural area with plenty of water and flowers. However, anyone can succeed in urban areas if they are willing to work hard. There are beekeepers throughout all 50 states. You can find beehives on rooftops, farms and balconies all over the country. There are many bee associations at all levels, including local, county, state and national. Although it is expensive, it is often only a few hundred dollars to set up the first bee colony. However, the joy and honey that it produces can make it worthwhile. Renting bees can offset some of the costs for farmers who need pollination.

Beekeeping is seasonal but requires different levels year-round. While winter will have fewer jobs that summer and spring, it will still require you to keep an eye on your bees and prepare for the next season.

In 1974, the Aebis family set the record for honey harvested per colony at 404 pounds (183 kg). We all know that records can be broken!

Chapter 2: Beekeeper Business

Why would anyone want thousands of venomous bugs to be a part of their lives for honey or fun? Why is beekeeping and apiculture so attractive to so many people?

Honey is the short answer. Honeybees don't have to be the only insect that makes honey. It is also made by some types of wasps. People don't keep honey bees because they are terrible tasting and difficult to work with.

Honey has been loved by humans since prehistory. Honey is more than its taste. It has unique medicinal properties that make honey useful. Honey has been a symbol for wealth and royalty since it is difficult to find. It was also used in religious rituals to offer it to gods and goddesses. It is still used in religious rituals, notably those of Buddhism and other Afro-Cuban faiths.

Beekeeping is not just about honey. There are many other benefits to beekeeping. Beekeeping offers both tangible as well as intangible benefits. Beekeeping can help you to appreciate nature in all its glory. It's a way to look into what may seem like another universe and gain a deeper understanding of the complexity of life.

The Ancient World of Beekeeping

The first attempts at honey collection from beehives by humans were primitive and lacking in grace. One common way to collect honey from beehives was to find a colony in the wild and keep an eye on them until they produce honey at their peak. Then, you would smoke the hive to calm the bees, then open it up and remove the honey combs. This is a great way to collect honey, but it can be problematic because of many reasons.

The destruction of the hive makes it difficult to collect honey from the same colony again in the future. It also slows down honey production and forces bees to focus their efforts on finding a new place to nest and repopulate.

Honey shortage was due to the time required to find and monitor naturally occurring beehives. It was not a commodity that could be guaranteed to be readily available. It's also very dangerous to smash bee hives. Also, bear in mind that wild honey was likely being collected before fire. This would be an additional step in a complex process already.

The positive side is that ancient people discovered a way to smoke bees to calm them down and make it easier to rob them of honey. This is the only beekeeping technique that has survived into modern times.

The practice of collecting honey was a tradition that existed in ancient times. After centuries of observation, refinements to the art of keeping them were made. The process no longer resembled a robbery at a convenience store. The bees were encouraged and supported to nest in artificially built hives, which allowed the beekeeping process be more beneficial to the hive as well as safer for those who tend to it.

These early hives were very common in Ancient Greece and Rome. These hives were usually made up of a hollow vessel like a vase, log, or log covered with removable slats. These structures were used as nesting areas for bees. They also allowed the beekeeper access to the slabs without disturbing the bees. These hives, also known as "Top Bar Hives", are still in use today in slightly modified forms.

Modern Beekeeping

In the 1800s, beekeeping was at its peak. The technological advances in scientific equipment have allowed us to get a better look at the anatomy and structure of the hive.

In 1852, a new type hive was invented. This improved understanding of bees' work in honey production led to a better hive. The Langstroth Hive is still in use today. It is named after Lorenzo Lorraine Langstroth (1810-1895), Philadelphia beekeeper.

This type of hive allows the beekeeper to let the bees go about their daily business without any interference. The honey can be collected in a nondestructive manner. Additionally, the hive's design allows for both routine health checks and maintenance without disturbing the bees that live inside. This hive allows beekeepers to collect more honey from the hive than they can from a wild or top-bar hive.

Langstroth hives are still being used today, and they are a popular choice among backyard beekeepers.

What would I do if I were a good beekeeper?

Before you get started with setting up beekeeping, there are a few things you should consider. To determine if you are a good beekeeper, ask yourself some questions.

Do I consider the environment an important issue? As one of the best ways to help the environment, beekeeping is a great option. The pollination of the food we eat and the trees that support the wildlife is the responsibility of the bees.

Are I a social person or not? You can join many beekeepers clubs, not only to socialize, but also for learning and sharing your experiences. These clubs are great for meeting other beekeepers who are interested in the same hobby and learning a lot from them.

Do you like feeling accomplished? Beekeeping is a time-consuming job that requires a lot of effort. You won't be able to quit once you start.

Are you interested in making your own honey and beeswax products? It is truly of the highest quality. The products on the market today that claim to be beeswax are petroleum-based waxes cannot replace it. Bees spend their time collecting nectar in order to make the best honey and wax. This cannot be done mechanically.

Are I able to perform scheduled maintenance on the bees? To ensure that they are healthy and continue to produce, bee colonies require regular maintenance.

Do you like to be surprised? It all works together, and the bees are happy, and the nectar will flow freely. You will be amazed at the amazing things that a bee colony produces.

Are you a person who likes making decisions? Maintaining a colony means you must make quick decisions. These are hands-on decisions

Are you looking for a rewarding hobby? It is not easy work but the rewards are great at harvest time. It's mentally rewarding when all your hard work is rewarded.

Things to consider before keeping bees

Once you've answered the above questions with a positive answer, it's time for you to think about beekeeping in more detail. Before you jump, you should carefully consider the following:

Is it possible to keep bees in your community? Do your research and contact the local government. You might need a permit or register. You may not find any restrictions, while others might make it illegal. Not all places are bee-friendly, so make sure to double-check. You could be fined a lot if you do not. Brooklyn woman was fined $2000 for keeping two bees on her roof before it was legal. You must get permission from your landlord/landlady/agency to keep them if you live in a rented apartment.

Learn everything you can about beekeeping. You need to learn more about beekeeping than just the equipment. Without understanding their behavior, you won't be able to raise a colony. Although you won't spend as much time caring for your bees than you would with a dog, they still require attention. You need to learn about basic care. Don't buy them just because they're trendy. There are many resources: clubs, libraries, and YouTube videos that can help.

Join a beekeepers' club. Take a class in beekeeping if you feel the need. These classes are offered by many clubs and can be paired with mentors to assist you.

Consider this: Why do you want bees to continue living in your area? Honey, honey-producing products, and helping out with the current bee crisis are just a few of the many reasons people keep bees.

Are you afraid of getting stung? No matter how careful you are or how protective your clothing, you can still get stung. Is that okay?

This leads us to the next question: Are you allergic to bee-stings? This may not be the right hobby for you. You can visit an allergist to have your allergy tested. You can still care for bees if you have only mild reactions. Just make sure you always have an Ep-pen and a mobile phone. You might consider giving this a miss if your reaction is more severe.

Ask your neighbors what their opinions are about beekeeping. Tell them why you want bees, and ask if you can keep hives in their yard. People tend to associate bees only with their stings. They don't realize the many benefits they provide or the great things they can do. A deal sweetener could be offered in the form of honey or candles made from beeswax. Although it might seem unnecessary, they might not realize that you did this later on if they complain.

You're going to keep your bees where you want. Is there a place that would be suitable for your hives? The hives should be placed on level ground and exposed to the sun during the day. They must also be protected from strong winds. They should also be easily accessible throughout the year. Honeybees can fly up to 5 miles to obtain the nectar and pollen they need, but it is best to give them an area where they can easily forage. You should also ensure that they have access to water, or are located near water sources.

Can you lift at least 25 pounds? That is quite a small amount! Honey-laden hives can be very heavy so you'll need to have some physical strength. You might consider hives that are lighter or don't require as much lifting if this is a problem. You can also ask your friends to help empty the hives.

Do you have the ability to reach the hives all year? You might consider hiring another beekeeper to assist you and look after your hives.

Is it something you can afford? This hobby is not cheap to start. A hive will cost you $300, and bees $100 in the first year. It is better to buy two hives so that you can double the cost. Make sure you do your calculations first to determine if it is possible to afford the initial cost. Maintenance won't be expensive once the hives have been up and running.

Do you think you can do it all on your own or will you need help? It might surprise you at the number of people who offer to help you when you share your knowledge so they can also learn. It is a good idea to live in an area that encourages beekeeping. The more people learn about it the more they will want to start their own hives. Sometimes you might be lucky enough to find a neighbor who is willing to help. This makes it easier and less stressful for you. When harvesting is coming up, it's helpful to have someone to help. You can also share equipment with a cooperative to reduce the cost for everyone.

What's next?

Now that you have done your research, determined that you can afford it, and are ready to put in the work, it is time to get started.

Before you start beekeeping, get everything set up. Set up your hive with all the equipment and lace. Reserve an area for extracting bees. It will make it easier to get your bees into a hive already established.

Order your bees. Order them now or wait until February to place your order. Most apiaries sell out quickly. You will not get any honey if you order them late or you won't be able to harvest the full year. You will lose a year if you don't leave honey in the hives for the bees to eat.

You will need to create a schedule for taking care of your bees. Everyone is different. It is important to determine how often you will inspect the hives and check for any signs of disease. You will be using chemical medications or natural methods to care for them. You must ensure they have fresh water whenever they need it. Your neighbor doesn't want your bee colony hanging out at his tap.

You should be prepared for busy times. There will be times during the season when you will be overwhelmed by bees. These are usually the swarm season or honey harvest season. The bees won't swarm the first year of being in a hive.

However, after that, they will. This natural process occurs in spring. Splitting up bees is a natural process. Half of them will leave the hive to search for another one. If the neighbors become scared by a low-flying swarm of bees, you will have to deal with them. Be prepared to take the swarms with you. Honey harvest is next on the agenda. You can expect to spend many hours gathering and preparing frames for extracting honey.

Enjoy your hobby and enjoy your bees. You can learn a lot from their behavior and intelligence about how to care for them. This old saying is very true.

"Remember, you don't keep bees; they keep you."

Equipment

A place to keep bees is the most important part of beekeeping. For novice beekeepers, a top bar hive can be a great choice. They are typically less expensive, easier to maintain and smaller than the Langstroth. They are ideal for urban and suburban applications that don't have enough space. If you are familiar with building tools and basic construction, a top bar hive can be constructed on your own.

A top bar hive can produce less honey, and maintaining the bees and collecting honey is more difficult than if it were a Langstroth.

Place your hive in a way that doesn't create unnecessary chaos for yourself or others. Worker bees need to fly every morning. You must ensure that nothing is in their path. If possible, elevate your hive so the bees can conduct their business a few feet above you (and your neighbor). Access to sunlight is important. Provide adequate protection for the hive from rain and other curious honey-loving animals.

Once you have set up your hive, you will need to locate a source for bees. Bees are usually sent by mail in the spring. These bees are known as "Package Bees". To find the best source of bees, you will need to do some research.

It is crucial to make sure that your bees are available in spring. This is when your backyard hive will be in its best condition.

Follow the instructions provided when introducing your bees into the hive. The queen bee usually arrives in a separate container. It must be placed first in the hive. Next, the drones and workers must be introduced into the hive. They need to get used to her before they can start making honey.

In the event that you do accidentally upset bees, you'll need a way to protect your self. Remember that opening the hive to check on the bees, or to collect honey for your morning tea is essentially ripping off the roof of their home, spying on them, and then disappearing.

The bees can get a bit agitated when this happens on a frequent basis. Although beekeeping rarely results in bee sting death, it is a good idea take preventative measures to avoid being stung. It is best to accept the fact that you will be stung at some point. If the beekeeper is mad, the rule of thumb is that it's their fault. Either the beekeeper is too inflexible or intrusive with their methods, oder something is wrong in the hive (diseases, wasp invasions, etc.). These are things that the beekeeper needs to be aware.

Smoke is the best friend of beekeepers and has always been. The bees' ability to communicate is temporarily disrupted by smoking. This happens because they are unable to smell their pheromones. They become confused and docile and are less likely to bother you while they check on their progress.

A "smoker" is a common way to smoke. It is a simple metal can that attaches to a set bellows. Smoke is made by burning pine needles, or other non-toxic materials that smolder. The smoke is kept at a cool temperature by smoldering, which will calm the bees and not make them sing.

A full English-style bee suit with a veiled pith cap and a bee suit may seem a little extreme to some. Protective gloves and a veiled cap should be sufficient to protect against bee stings. A full suit's usefulness is not only its practicality, but also the psychological protection it provides that allows you to remain calm and steady as you enter the hive. Many beekeepers begin with a full suit, and then reduce the amount of clothing as they become more comfortable and skilled.

Chapter 3: Introduction to the Bees and The Hive

Your bees now have a home to call their own. You will be excited to bring the bees inside the bee box as a novice beekeeper. However, it is not enough just to collect the bees from the beebox. This is a process that requires careful consideration. This chapter will provide additional information about beginner beekeeping and the steps involved.

You will need to first learn about where and how to contact honeybee sellers. Next, you will need to know how to bring the bees home. You will love getting the bees.

Where can I purchase bees?

Association of local beekeepers

There are many beekeeping organizations in different areas that deal with bee-related problems. These associations encourage beekeeping among their members and engage bee-related, economic activities. These associations ensure bees are sold at an affordable price and maintain bee health. If you're lucky enough to find one, you can purchase bees from them.

You will get some benefits from buying from these bee associations:

Learn about the local bee guidelines

Local laws and regulations pertaining to bees

Their treatment is covered by laws

Economic viability of bees

Care for bees, especially as a beginner

Information on how to feed them, and where to find the best food

You'll also get to meet other beekeepers, and gain valuable information from their experiences.

Buy from another beekeeper

There are many reasons people keep bees. Some keep them for crop pollination while others keep them for their honey production. If you are looking for your first bees, a seller who has bees for sale might be able to help you. This seller will have taken care of the bees and raised them to the required standard. It is better to buy bees locally than import them.

You should ensure that the beekeeper you buy bees from is free of disease, parasites and pests. You don't want mites or hive beetles to bother your new colony. The losses that such pests can cause will hinder your entry into beekeeping.

Buy packaged bees

Packaged bees are packaged in a box that contains a queen bee. This bee source is popular in the USA because it can be shipped by mail. These are the most popular methods used by beginner beekeepers to get started. The typical package contains bees in a container with screens and the queen in a cage. The container will also contain bee food.

Install the hive once you have received your package. Do this in the late afternoon or evening so that the bees are able to enter the hive. After removing the queen bee from its cage, open the hive and the colony will happily accept it.

You should be cautious with packaged honey because there are some fake sellers. You should ensure you verify the seller's credibility, have a consultation and ask the right questions prior to purchasing. You should ensure that the package contains no honeybee diseases or mites.

Nuc hive honey bees

A small nucleus hive is another great way to get bees. It has just a few frames of honey and brood. The hive includes a queen and several hundred bees. Most honeybee sellers will also have this option available. The nuc box is typically purchased in spring and placed in a regular beehive box.

The nuc boxes will allow the bees to expand into the beehive boxes, and can be removed later. The nucleus hives can be a great way to jump-start your honeybee colony. The bees will also benefit from honey in the hive as it will provide their first food source before they begin rummaging for nectar or getting pollen.

The nuc hive has another advantage: It is a fast way to start a new colony. The bees are able to quickly get to know each other with this method. The queen bee is also easy to get to know: the frames and combs are already drawn.

How do I reach honeybee producers?

There are many ways to reach honeybee sellers. Local sellers or the beekeepers' group can help you buy honeybees. It is best to meet face-to-face with them so that you can inspect the bees. You can also contact the sellers via email or telephone.

If you are unable to meet face-to-face, it is important that you do your research to verify the authenticity of the supplier. There are many great bee sellers that you can find in ads and newspapers. Ask your friends and beekeepers if they have ever bought from any of these bee sellers.

There are many resources online that will help you find local bee suppliers. Don't rely solely on what you find online. Always do your research on each supplier and, if necessary, start small with batches.

What are the factors to consider when buying bees?

Be a true bee seller

Before buying, verify the credentials of the seller. A good seller will have experience with beekeeping. You will be able to find out more about the bee seller by reading reviews online and speaking with previous customers.

Avoid beekeepers who have negative reviews and don't keep their promises. Poor handling of bees can make them more vulnerable to diseases and pests.

Local beekeepers can help you buy bees from established colonies

Most likely, you already know that bees live in colonies and are social insects. Their survival is dependent on each other's safety. If you purchase bees, it will be easier for them to reproduce quickly and help with their survival.

Place at least two colonies in separate hives. Double placement allows you to still access the correct hive in case one is damaged.

Ask beekeepers if they are able to sell bees in Nuc hives when you purchase bees to start a colony. This will allow you to have a healthier colony and bees.

To ensure that your shipment boxes are free from mites, always check them.

Before you put the bees in the hive, make sure they are free of bugs and ants. These predators feed on the honey and larvae of bees. The honey can be eaten by ants and bugs, which could lead to the colony moving.

Beekeepers who have bee colonies infested with mites, beetles, ants and other diseases should not be selling them. For everyone's safety, honey only from healthy colonies should be sold.

Many regions have regulations regarding honeybee importation and sale. If an area is clean of varroa mites it would not be illegal to import bees from colonies infested with them. Although you can control the pests in these situations, it is best to prevent the problem from ever happening.

Purchase bees in starter or nucleus hives

The nucleus is made up of queen and worker bees. It can take around 4-5 frames to complete. This type ensures that your colony has a queen, workers and drone. The nucleus, unlike packaged bees has a much higher chance of survival in any beehive.

Things to Consider Before You Buy Bees

The next step in your honeybeekeeping journey is to get the colony. Here are some things you should keep in mind:

Learn about the different kinds of bees

Before you start keeping bees, it is important to know what types of bees are available for beginners. There are many factors that affect the temperament of bees, including their resistance to disease, foraging ability, parasite control, and foraging power. Although it is beneficial to have bees that live in your area, this will help them adapt better. However, you don't need to be concerned about all bee species. Ask other beekeepers around your area to help you determine the best bee species for your environment.

Get your housing ready

The beehive was discussed in the previous chapter. Make sure to inspect it again to make sure it is safe for bees. Is it easy to use? Can it expand? Is it able to keep the bees warm in winter? In case of an emergency, you might also want to keep spare parts in stock.

Get beekeeping equipment

When you are preparing to bring the bees in, be sure to have all your beekeeping equipment nearby. Beekeeping is more than just about managing bees. You will need the necessary tools to manage them properly, including hive tools and bee feeders as well as mite treatments. Parasite treatments are also available.

Chapter Seven will provide more information about how to control pests in beehives.

The installation colony of honeybees needs to be fed

You should first get bee food before you bring your bees into the hive. In Chapter Four I will cover the topic of feeding bees, and what kind of food you should choose. But let's first look at the concept.

Sugar and pollen patties are two of the most important sources of bee feeds. The sugar is dissolved into water, and the pollen patties are given to the bees in their original form. These bee feeds can be purchased in bulk and stored until bees begin to produce their own food. A healthy colony of bees will settle in the hive quicker, become stronger, and produce higher-quality honey.

Beehive location

Your beekeeping activities are influenced by your location. It is the most important factor that determines the yields you receive from the bees. You should consider the proximity of your home, as well as a safe area. To ensure that bees don't sting you, please avoid areas where there is a lot more traffic.

Respect local laws and regulations governing beekeeping

You must be aware that different countries and states have their own laws regarding beekeeping. You can avoid problems that could lead to your beekeeping operation being shut down by complying with the laws.

Can I catch bees instead of buying them?

If you don't have the funds to purchase bees, you can capture a swarm and put them in a beehive. Sometimes, a honeybee colony may split into several swarms. This will allow for the establishment of new colonies. Beekeepers can capture the swarms to use in their honeybee colonies.

Because of their genetic variation, wild swarms are a favorite among beekeepers. These variations can also add unwanted characteristics to your honey pool, which will impact the temperament of the bees.

It is difficult to get a swarm of honeybees. Even experienced beekeepers have difficulty with it. While traps may be set to capture the swarm, in some cases you might not get any. Because swarms are unpredictable, this makes it difficult to find a swarm. This should not be done as a novice.

You can start to use trap boxes that look like trees once you are more experienced beekeeper. The hollows should be small enough to attract bees. However, they shouldn't be too large or heavy for them to carry when they capture a swarm.

Here are some steps to help you place your bees in a hive

Make sure you have the following tools before placing your bees into the colony: the hive stand and bottom board, one frame (for a 10-frame hive), inner cover and entrance cover, as well as a feeder.

Step 1.

As soon as the bees arrive, get them and check that they are still alive. If there are dead bees at the bottom of the container, it is a sign that there is a problem. Please call the supplier. Install the bees as soon after they arrive. If you prefer to wait a few days, you can put them in a cool, but not cold, place. You should place them in a dark space such as a garage or basement. Keep an eye on them. If you are able to install the bees right away, do so! This would be the best way to go.

You will also need a hive tool and a lit smoker, as well as smoker fuel and lighter.

Step 2

You can now install the bees in the afternoon if you make sure the smoker is lit. The smoker isn't usually used to install packaged bees but it is a good habit to start with inspections of your hives. Depending on how they were obtained, open the lid and remove the feeder jar. The tab for the queen cage is still in place. To prevent the bees from escaping, remove the queen cage and place it in a new lid. Take 3-4 frames out of your hive. If the queen bees are clustered on the queen cage, remove them and brush them off.

Step three

To make sure that the queen is healthy and well, inspect her legs and watch her walk. If possible, count the legs of the queen to make sure they are straight. If the queen appears to be in pain or injured, put her back into the cage and return your bees to the seller. You can place the queen in the hive if she is healthy.

Fourth step

There are three ways to place the queen in the beehive:

The rubber band method involves putting a rubber band around a frame in the hive, and placing the queen's cage between it and the foundation. The cage can be placed vertically, just behind the rubberband.

Next, replace the frame by the queen cage in the middle of the hive. The queen should always be in the center.

The rim board method involves placing the queen's cage on top and placing the rimboard around it. Once the cork has been removed from the queen's box, place the cage on top of the bars so that it is in the middle of the hive. Place the rimboard on the hive so that the queen can be seen.

You can also use a drawn brush to push the queen's box into the wax and stick it to the comb. This is a great method to use when it's warm outside. Make sure the candy end is visible: After the queen's cage has been secured, you can place another frame on top of it. Once the queen bee is installed, it's time to install the bees.

Step five

A few bees can be shaken on top of the queen to release "Nasonov Phenolone", which tells other bees about the presence of the queen. You should get as many bees as possible into the hive from which you received the frames. Don't worry if they don't all get out. The rest of the bees will find their way back to the hive.

Replace the frames, but don't push or crush them from the top. Let them fall in place gently. If you used the "rim board" method, place the rimboard on the hive first, then cover the hive with the inner cover. When replacing the covers, brush the bees out of the way. Keep the original container or packaging you used to transport the bees near the hive. This will allow the colony to be found by bees that have been left behind in the transfer process.

Sixth step

While I will be discussing feeding in more detail in the next chapter of this book, it is important to know the basics of the first stages of feeding the bees after they are installed. To feed the bees honey, beebread or pollen they can be fed through a feeder. They will then need to be fed until they are able to make their food, build a home, and start to wax.

Keep the feeder full. You can also feed them 1:1 syrup (your local honey shop will know this). This will keep them on the ten frames. However, don't give them too much.

Step seven

Do not touch the hive after installing your bee packages. They might think you are disturbing the queen and decide to kill you instead of accepting it. The bees will fly around if the feeder is full. This is normal as they are still learning their new surroundings.

This is the time to look for bees that have been carrying dead bees. It is an indicator that the colony has thriving. You should also be on the lookout for bees that bring in pollen, which is necessary to feed the hive. This is a sign that the queen is coming soon.

Step eight

After one week, get dressed up and start smoking the bees. Take a few puffs at the entrance, and then take a deep breath under the cover. Check the queen's cage to make sure she is freed. If she has been released, it is time for you to take out the queen's cage. If not, check to make sure she is still alive. The queen should be released from her cage by removing the cork from its candy end. After removing the cork from the candy end, place the cage back into the hive, and close it again. After a few days check the hive again for eggs. Make sure the queen is laying eggs.

How to Identify the Queen Bee

You may have noticed that I mentioned the queen honeybee throughout this chapter. However, if your bees were purchased from a wholesaler or packaged service, they might be able to separate her for you. If you bought your bees via a different source, it may be more difficult to identify her.

The queen bee, also known as the worker or drone bee, is the leader of the colony and mother to most bees. The queen will ensure that the hive thrives if she is healthy. If the queen becomes old and dies, the queen will pass away. To maintain your hive you will need to distinguish the queen from other bees. The steps below will help.

The queen bee is the longest and narrowest.

Honeybees all have a flat abdomen. But queen bees' belly is pointed.

Queens are more visible than other bees because they have their legs spread out, while others have their legs below their bodies.

One queen is allowed to live in a hive. If you find multiple queens, lift them up to examine the middle of their bodies. A magnifying glass can be used to inspect the queen's body. She will have smooth, un-barbed stingers.

The queen can be found by gently removing the hive frames. Because she lays all of the eggs, the queen will be near you. Be careful not to injure or even kill the queen.

The queen can also be identified by her movements. Drone and worker bees will never move in her direction, but they will cluster together where she is.

The queen is taken care of by the rest of the hive. You can easily spot her by looking for bees that aren't doing anything other than eating.

Once you have identified her, color her. There is a way for beekeepers to do this. Based on the number of years left, beekeepers use specific colors for identifying the queen. You can identify the queen and protect your hive by marking her.

White paint should be used on the queen's thorax if she has between 6 and 12 years before her reign ends. Yellow paint is recommended for queens who have between 2-7 years to her reign. For 3-8 years, you can use red paint, while for 4-9 years you can use green, while for 5-10 years you can use blue paint. You don't want her to be hurt by the act of removing, marking, or putting her back.

Depending on where you live, beekeeping becomes more popular. However, accessing healthy bees can be difficult. Local beekeepers' groups and beekeepers are a great way to get bees. Locally bred bees are more adaptable to local environments than imported bees so it is worth buying from them.

Before placing bees into the hive, please obtain approval from your local government. Once your bees have been placed in their hives, you will need to begin the process of caring for them. The first step is feeding. In the next section you'll learn how to care for and feed bees.

Chapter 4: Beekeeping Practice

These are some of the most important characteristics of bees

Honey bees take their brushes from the top. They don't often annex their brushes to the base or to inclining walls. The honey bee space is the fixed partition between the bush (the honeybee space). Brush will finish off any space that is more important than the honeybee space.

It is easy to see that adaptable brush beekeeping is huge. This honey bee space, which is similar to the brush scattering (for example the detachment between focal points of adjacent brush) is constantly fixed for settlements by a similar honey bee race. However, it is still relatively changing between tracks. It will generally be stated that the smaller the honeybee species, the more honey bee space, brush scattering and cell size, the home volume, will all be affected. Blooming plants are a good source of honey and residue for honey bees.

By spotting rummage honeybees on plants, you can easily determine if they are reliable sources.

Similar to honey bees, they also need water close by that they can bring into their hives.

These essential ingredients should be available to the beekeeper in case they aren't. The majority of the time, the area will either pass on or disappear.

Base-bound honey bees can only fly to their own hive. They will fly back to their hive every day, regardless of whether the colony is moved.

Honey bees are unable to travel to other states because of their home smell. N.B. N.B.

Beekeeping

Beekeepers should be aware of the way honey bees react to specific aromas such as sweat, alcohol, chemical, and smell. These intense smells are not likely to cause stings so make sure to inspect the honeybee territory and do away with any animals that may be in close proximity to the honey bees. Honey bees can become trapped in hair or woollen clothing. Therefore, it is best to cover your head with smooth clothing. When honey bees are very intense, they will always choose dull colors first. Use the lightest concealing possible. It is also a good idea to wear light clothing when you are working in hot airs. You should ensure that there is some smoke nearby when you open the hives. For honey bees from South America and Africa, you will need to blow smoke into the entrance. After that, lift the spread and blow some smoke into your hive.

Finally, close the hive for a short time (one minute). You must always ensure that there is enough fuel to light the current smoker. Viable honey bees can be disturbed by vibrating objects and machines. You can keep your apiary in a healthy state by carefully inspecting the area. The honey bees can be stimulated by weeding and cutting the grass with a sickle. (See similarly: Setting up an apiary). All activities should be completed with slow improvements. The honey bees are receptive to rapid advancements. You can be stung regardless of whether you are allergic. It is important to avoid hammering on the colony. You should first kill the honeybee that has stung and then, a few days later, use a needle or fingernail to remove the sting. The bites can cause honey bees to develop right away when you first start keeping them. After several bites, your reaction will be less. If you are unable to control your reaction to honey bee stings (sweat, shakinginess), then it is time to get rid of honey bees. This reaction is rare (1 in 5000), but if it does occur, consult an expert immediately.

Fixed comb beehives

You can make these using an empty log or a wooden box, a wooden pot, an earthen container, or a metal holder. From the top, the honey bees take up all available space. The bushes are attached to the sides and top of the hive so you can't remove them on their own. You can expel the honey by removing one mass from the colony, and then breaking up or removing the honeycombs.

Fixed comb hives offer favorable circumstances:

These are easy to make and very simple.

They are occasionally grieved by ants or raccoons from time to time. The hives are usually hung up in trees or covered with a layer mud.

A beekeeper doesn't constantly disturb the colony.

Fixed brush hives that are better have at least one removable sheet, which allows for inspection from at least one side of the hive. This allows you to remove only honey-containing brushes and monitor the colony's progress. The province will build new brushes and leave the brood-rich brushes in the hive. This allows the colony to continue its development. The Kenya box hive uses a similar standard.

Paraffin tins could also be used as hives. The other side is cut off. The tins should be protected from excessive cooling or heating by wrapping straw or other comparable material around them.

Colonies equipped with movable hair combs

These colonies are used in tropical Africa to A.mellifera (Kenya and Tanzania, Botswana; Ghana), as well as in Asia for A.cerana. (Nepal, Vietnam).

Kenya top bar hive. Bars of fixed width are used to verify the box is a long, trough-shaped one. As long as the spacing between the sticks is correct, round sticks can be used in the same way. You can research a small portion of the brush area, while the rest of it is verified. This is a benefit. Honey bees can develop a brush on any of the bars if someone melts a touch of brush 2 to 3 centimetres from the underside or dips the bottom in fluid wax. This type of hive has one essential component. The honey bees do not add brushes to the slanting side-dividers. A honey bee space is without left. The brush isolating should be exactly the same as the centre division of top bars. This is applicable to all honey bee races. One of the most fascinating enhancements you can find is spacings for certain honey bee races, A.mellifera or A.cerana.

If you are using round sticks, they should be placed at the same divisions (evaluating from the point where the sticks meet). You can find the division between honey bees by looking at the brushes from a settled that has a fixed brush home.

A baseboard, two side partitions, and a front and rear divider make up the hive. You can make the baseboard wider than what is shown in the drawing. This would allow the distending fragment to be used for a flight load-up (runway) by the honeybees flying in. To fill in the flight sections, two cuts measuring 1x15 cm were made in front of the divider.

You can make the spread using any material that provides palatable protection from light, sun, and deluge. The bars' underside should be shaped into a V shape. The bars should be the correct width. If sticks are used, it is important that they are placed at the perfect distance from each other using techniques such as scattering bars or nails. The rods or bars should measure 48 cm in length.

The hive can be delayed between two posts or trees with stiff wire. This will protect the colony from termites, ants and other enemies. To protect the province from excessive heat, it should be painted white. The wood could be treated with an additional substance, but not with bug spray. You can find out more here.

The Tanzanian top bar honey bee

This type is found in many parts of Africa, such as Botswana. It is simpler to make than the Kenyan type. This type is for honey bees that don't connect the brushes to the side-dividers.

You can also use rectangular top bars instead of V-shaped ones. A hardboard strip is placed at the point of convergence for each top bar. This piece should be approximately 1 cm wide and can be used to make a starter for honey bees. You can make top bar hives using 2 cm of thick wood. An inexpensive way to improve your dairy animals' health is to use a cardboard box that can hold their waste, earth, or some combination. You can also support the cardboard box hive with wooden sticks. You can also make a case from straight sticks, wired and secured with wire.

Top bar hives have many advantages over fixed brush hives:

Every brush can be removed from the colony. This allows you to monitor the state's improvement.

Honey-containing brushes can be removed without causing damage to the brood home.

Honey is more natural when there is no brood.

To isolate the brood brush and honeycombs on one side, or both, you can use a few queen excluders. Queen excluders are boards with specific openings that allow the working drones to pass but not the queen. This allows you to obtain honeycomb with some pollen, but no brood. The queen cannot lay eggs in them.

These hives are superior to hives that have outlines for the brushes because of their benefits:

You can also make them with cheaper and locally available materials.

There are only two measures that matter:

The length of the standard top bar to allow for the possibility of changing the top bars between colonies and within the hive.

The top bar's width must equal the honey bees regular brush-dispersing, with the goal of allowing them to make brushes directly underneath each bar.

The crude wax should not be washed if the brushes are used only once. This is because the wax production is very high.

To remove honey from the brushes, you don't need a centrifugal honey extractor.

Frames for bees

There are many types, such as WBC, Langstroth and Dadant. Simplex hive is another. This is not the intention to go through all of them in detail. This can be explained by using the Langstroth and East African Long hives as models for the development of this type of colony. Honey bees attach their brushes to wax sheet starters (or establishment sheets), which are fixed in wooden edges. An establishment sheet is a layer of beeswax between 2 and 3 mm thick, slightly less than the inward proportions. It has been used to engrave a hexagonal example (the condition of the base of a worker cells) with a press. An establishment sheet is attached vertically at the edge. The cell dividers are made by honey bees using a flat plane that is parallel to the establishment sheet.

Preferences of a beehive with frames

Brushes with honey can effectively isolate combs with brood.

The edges of the brushes are securely secured so that the bushes can quickly be tended to without breaking. The hives can easily be moved with the brushes not being cut.

The honey can be quickly removed from the edges using a diffusive honey extractor. After that, the brushes can be reused.

This gives you a lot of flexibility if you're primarily interested in creating honey and not beeswax. Honey bees must produce less wax so that they can use all of their vitality to care for the brood as well as the variety of food.

There are two advantages to beehives that come with frames:

Langstroth hive (North America, South America, Africa and Australia).

The East African Longhive (Uganda).

Chapter 5: Using Nectar Substitutes

The nectar is produced by plants through a glandular secretion called nectar. It can contain very little vitamins, proteins and other nutrients.

There are two ways that bees use nectar. One is to substitute for water and the other is to ripen nectar to be used for different purposes.

The colony should be inspected approximately every ten days in early and late spring.

In the spring beekeepers will always feed the bees a pollen substitute and if the bees need to be fed sugar syrup.The sugar syrups fed early in the season are used for brood rearing.Feeding sugar usually stimulates egg laying and the syrup is usually a "light" syrup mixed with 1 part sugar and 1 par water.A heavy syrup, a mixture of 2 parts sugar and 1 part water, is fed late in the season to ensure adequate winter food supplies.They are stored as ripened syrup.If a medicated treatment is needed in the fall, feed for weight first, and then top off the colony with medicated syrup.There are beekeepers who use high fructose corn syrup to feed their bees, but they do not usually dilute the syrup regardless of the season.There are some levels of hydroxymethylfurfural (HMF) that will increase over time, especially with heat.HMG is toxic to honeybees at high enough concentrations.

Each colony should be given its own syrup.

Using Pollen Substitutes

The pollen provides honeybees with a good source of vitamins, minerals, and protein.

Brewer's yeast is the main ingredient in making a pollen replacement. However, most beekeepers don't use this because it is expensive.

The beekeeper can feed the bees a pollen replacement to increase their brood production.

The pollen substitutes are not liked by bees.

Chapter 6: Bee Equipment & Bees

You should now understand the reasons anyone would be interested beekeeping. This chapter will cover everything you need to know about beehives and beekeeping equipment. Learn about beekeeping tools and how to buy suitable clothing.

Beehive Structure

It is recommended that you start with at most two beehives when starting a beehive. This way, if one colony dies, you can take the bees from the other and ensure their survival. Although a beehive may seem complicated at first glance, it is actually very simple. It is almost like looking at boxes stacked on top of each other with a lid to stop rainwater entering the hive.

These are the elements that make up a beehive.

A hive stand is a stand that you place on the ground. The excessive dampness can harm the health of bees and cause them to die.

Floor - The floor is the next item after the hive stand. It is actually a mesh which is attached in a wooden frame. Made from stainless steel, it will support the brood box as well as facilitate ventilation. However, the floor does more than just protect the bees in cold weather.

Brood box: This is where the queen will lay eggs. If your colony is growing too fast, it is advisable to add another brood container. This will reduce the chance of them swarming.

Queen excluder – Located on the broodbox, the queen excluder separates the queen from the rest of colony. However, the flat grid allows the worker bees access to the higher boxes and stores the honey. The queen cannot pass through the grid because she is too large.

Honey supers are found on the top of the brood room and serve the purpose of storing honey (3/4 containers are recommended).

Frames - These are found inside the boxes. They can be made from either wood or plastic. Beginners often choose wooden frames because they are simpler to use. However, experienced honey extractors prefer plastic frames. Hoffman is the best frame for the brood chamber (10 in each box), and Manley (8-9 in each case) are the best frames for honey supers.

Frame feeders – These are also made of plastic and serve the purpose to feed the bees sugar syrup (if needed).

Foundation wax - This is for beehives with no bees. The wax sheet will fit into the frame and have a similar shape as the honey comb.

Crown board - This is not an essential part of the beehive. It is located at the topmost box, just under the protective lid.

The lid - There are many options, but the flat or telescopic lid is the best. It also prevents rainwater from entering the beehive. If you live in an area that is hot, you can have the lid painted white.

You should also consider buying the following equipment/tools:

Even though it's a simple metal bar, the hive tool is one of the most important pieces of equipment for beekeeping. It will allow you to access areas otherwise impossible to reach because the bees have used propolis to glue them together.

Smoker - It is well-known that smoke can calm agitated bees. All it takes to achieve this effect is a small amount of smoke at the entrance of your hive. You should make sure your smoker has a protective grid. This will allow you to use the device without worrying about burns or gloves.

Divide board - If you're interested in splitting up your hive into two distinct ones, this tool is for you. The division board can be used when you start with a hive with two queens. This will allow you to maximize honey production.

Clothing

To reduce the chance of getting stung, it is advisable to purchase a complete body suit if you are just starting out in this field. You can use a hood without the veil or a full-body suit as you get more experience. It is best to wear a suit that has a zippered hood. This makes it easier to remove the suit when you're done with bee work. If you work with more colonies or bees that are aggressive, a full suit can make a great investment. You will need gloves, made of leather, to protect yourself from being stung. Gumboots provide adequate protection for the feet.

Bees

It is best to buy beehives from a local beekeeper. This will ensure that your beehives are installed correctly. However, this is a good way to get started.

Hiving a swarm is an alternative, but it's still possible. You can get a swarm of bees from a beekeeper, and ask for his/her advice on how to properly install them. A nucleus of honey bees can also be purchased. This usually comes in a small box that contains several frames (e.g., 1 brood frame and 1-2 honey frames), as well as 1 honey comb frame. The laying queen is also found in the nucleus. This possibility has the advantage that the brood is already ready to emerge and can contribute to the survival and growth of the colony.

You can also buy bees as a package. A package that contains regular bees can contain between 8000 to 12000. It has four sides made of wood and two screens (anterior, posterior). You will also find a queen laying in separate cage made from wood. If you choose to opt for this alternative, be ready to wait three more weeks until the brood emerges.

If you're wondering when is the best time to buy bees, it is spring. This will allow you to witness the whole development of the colony, from a small nucleus or colony to a vibrant beehive ready to produce delicious honey and other products.

Chapter 7: Beekeeping Worldwide

Honey and beeswax are produced in many regions around the world for both therapeutic and nutritional purposes. There are many beekeeping businesses in the United Kingdom, including those in England, America, Africa and Asia. While beekeeping is a tradition that has existed for centuries, modern beekeeping was developed in Europe. There are many ways that beekeeping can be taught and practiced today. It's fascinating to see how diverse cultures can be. Each country has its own teaching methods and training techniques. Beekeeping isn't just a job in some cultures. It is a social activity.

Subculture and way to live.

Honey is used in many places around the globe, not only as a sweetener and food, but also for religious ceremonies and celebrations. Honey is used by some cultures as an ingredient in special foods and drinks. Honey is also used in rituals that commemorate a holiday or event in their history. American honeybeekeepers primarily produce honey for mass production and shipment to supermarket chains nationwide, as well as export. Modern beekeeping techniques in America allow mass production in many areas, both within the state and outside of the country's consumption limits.

Many other countries don't use the same modernized system as the United States, which allows them to produce large quantities of honey. This efficiency allows America to produce enough honey each harvest season to supply the market and the economy until the next production season.

The bees are most active during the spring months, when honey and pollination begin. The mating season for bees begins in late March or early April. The spring brings new and reblooming plants to the beehives, which allows them to feed and pollinate abundantly. The hives themselves are easy to maintain, about an hour per day in peak season.

May to September depending on the region and weather conditions. The United States is the only country that supplies the majority of honey sold in American supermarkets. These stores and restaurants chains also ship honey overseas to sister or affiliated chains.

A honey harvest that is exceptional can yield sixty- to one-hundred lbs of honey depending on how big their beekeeping operation, the number of hives present and the happiness of their bees. Although honey is often purchased on a per-pound basis, this is not the only way sellers and buyers can negotiate. Buyers should consider the type and quality of honey produced by the bees, as well the health of the hive. The production of each batch of honey or hive products can vary depending on the amount of pollination and where the hive is relocated. It is sometimes just a matter what the bees eat, as different flora or vegetation can cause distinct tastes.

Parasites are the biggest problem for beekeepers during harvest season. To increase honey production, many beekeepers move their colonies from one season to the next. This is known as migratory bee movement. Seasonally moving hives is a great way for honey bees to have a fresh source of food to help them increase their honey production. Although moving hives can be tedious, it is worth it in the harvest season.

Chapter 8: The Sting in the Tail

This chapter will teach you how to avoid being stung and what to do if it happens.

Bee stings can be very unpleasant. These stings can cause severe pain and even lead to allergic reactions that can prove fatal. There are a few things you can do to make sure you don't get too worried.

Safety First

I don't recommend you having your own bee hive if you have been diagnosed with an allergy to bee stings. Being stung as a beekeeper is an inevitable part of the job.

Bees Don't Want To Sting You

Stinging someone is considered a death sentence by a honeybee. If the bee feels the need for protection, it is used only as a last resort. A bee that was bred to beekeeping is usually very gentle. You would need to be very careful to get them to sting.

How to Avoid Being Stung

Dress for the occasion: Wear trousers or jeans if you're going to inspect your hive. Also, wear a long-sleeve shirt and pants to protect your legs. Wear a veil to protect your face.

*Be sure to time it right. Think about it as going to the mall. The mall will be less crowded if you go in the morning, when everyone is at work. It can be chaotic if you go in the evening when everyone is off work. If possible, go to the hive between nine and five in the morning. The colony should be kept out of reach once it is dark.

*In fair and foul weather: Storms, inclement weather, and other severe weather conditions are reasons to avoid your hive. Bad weather is more likely for bees to be in your hive and they won't behave as normally.

*Keep your hands on the ground: Don't drop your frames when moving them. This will cause bees to become upset.

*Be patient and slow in your movements to avoid antagonizing bees.

*Do not whack bees. The hardest part about this hobby is the possibility that bees will start crawling on you. It can be nerve-wracking to see them crawling around your face the first few times. They are not making an aggressive move, but this isn't their intention. They simply want to be able to see the situation. You can brush them off if you feel the need.

*Good vibrations: Bees love consistency. If they feel a sudden vibration they will be ready to defend their hive. Loud noises are the same. Make sure there are no loud machinery nearby when you place the hive.

*Be clean. Bees and humans don't like B.O. This means that you should clean up before you meet your bees. If necessary, take a shower before you go to meet your bees.

*Lighter colours: It could also be a problem with the color of your clothes. Avoid darker colors as bees are more aggressive in lighter colors.

If You Do Get Stung

It is possible to be lucky and not get stung again. However, if this happens while you are at work in the hive you will need to remove the stinger and put smoke on the area. This will ensure that other bees do not pick up the scent and become alarmed.

Do not squeeze the stinger, as it can release more toxin into your body. Instead, scrape the stinger off. Cool water and a cloth can be used to reduce swelling. You can apply a cream to reduce the itching if you wish.

Anaphylactic shock is a reaction to being stung by bees for the first time. These include constricting airways, shortness of breath, swelling, a rapid, weak pulse, dizziness and low blood pressure[6]. Get medical attention immediately if this happens.

You can ask your doctor for an EpiPen if you have concerns. This is something I believe is a good precautionary measure until you can prove that you aren't allergic.

This is not something to be concerned about as very few people experience a serious reaction. A few stings can actually be beneficial because they can build your immunity to the toxin and make them less irritating.

Fast fact: To be infected with enough venom to kill your bees, as long as you're not allergic, you wouldn't have to be stung more than one thousand times.

This chapter teaches you:

*The weather and timing can have an impact on whether you get stung.

*Be aware that you should wear light-colored clothing to protect your body and a veil to approach your bees.

*Bees should be treated calmly as they are sensitive and can be irritated by loud noises, sudden movements, or body odor.

*Antihistamines, cool water and cool salt can reduce itching and pain caused by a sting.

You should be aware of signs and symptoms of anaphylactic shock, and seek immediate medical attention if you experience this reaction.

*Keep an EpiPen handy and treat your bees with respect.

The next chapter will teach you how to select the right place for your hive, and when it is best to place them.

Chapter 9: Understanding why your honey is different in color and flavor

Honey straight from the hive can have unique flavors depending on where it is grown and when it was harvested. These delicate combinations can be enjoyed in the same way as wine, with little context. Although the plastic bears that you can squeeze into your hand are labeled as honey, they actually contain a complex, homogenized drug. To maintain consistency in flavor, color and viscosity, something may be used to mark a specific season. This is then melted down to create a bland syrup that everyone has grown to love. The bland honey version is rarely a delight for the taste-buds. Honey has many qualities. These include color, texture, viscosity and flavor. It also crystallizes easily. These characteristics depend on the nectar collected by the bees in the hive.

FLOWER FORAGING

To attract insects, flowers contain nectar (sugar fluid) which attracts all kinds of insects. The sticky pollen from flowers sticks to insects' bodies when they search for nectar. The process of fertilization begins when the flower pollen and its stigma meet, which allows the plant to produce new seeds. Apis mellifera is the most common genus of honeybees. Although honeybees can also be produced by certain wasps, they don't store it in the same quantities as honeybees. Honeybees eat nectar directly from flowers, but they also bring pollen and nectar back to their hive. The pollen is stored in bee bread, which is rich in protein, fat, and other nutrients, and is fed to the brood. Honey is stored as honey during unfavorable conditions.

SWITCH NECTAR TO HONEY

Honeybees mix nectar and enzymes when they forage, and then store the nectar solution in a pouch called a honeystomach. The enzymes reduce the sugar into simpler forms that are more resistant to bacterial growth. These worker bees return to the hive to continue dehydrating the nectar/enzyme solutions by pushing it around in their mouths, and then depositing it in hexagonal cells that form the hive. The dehydration process is then continued by the house bees. To reduce the water level to 18 percent, the house bees spread their wings and cover the entire hive cell with wax. The honey is then called honey. This honey has remained pure and unfermented for years - even ancient Egyptian tombs were found to contain unspoiled honey!

THE HONEY IS TRACED BACK TO IT'S ROOTS

The season and availability of flowering plants in the area will determine how bees obtain their nectar and pollen. This all depends on the honey's texture and color, as well as how it smells. The honeybee is engrossed in pollen granules as it forages in the flora, which collects nectar. This pollen is not only a key ingredient in a hive's special features, but it is also traceable to the source of nectar that the honeybee visited during foraging.

A HONEY'S UNIQUE TASTE

There are many types of honey from different hives, with different flowers at the same time. You can also find "univariately", such as blueberry honey. Univariately honey is made by placing a hive at a place where one type of plant is blooming in excess within a radius of 3 miles (4 kilometers). Honeybees can fly up to 4km (3 miles) per hour to drill.

However, most of the nectar or pollen collected from honeybees will come from one plant. This will ensure that the honey has a distinctive flavor. These honeys are harvested right after the flowering of a particular plant, usually a flower. Univariate honey can still have a unique flavor depending on where it is grown and the growing season.

Sometimes, the appearance and taste of univariately honey can be correlated with the dominant plant it was foraged from. Wild blueberry honey, for example, has strong blueberry flavoring and even an indigo hue to it. Honey often doesn't have a clear correlation with the flowers it has harvested. Buckwheat honey is not a true buckwheat flavor. It's a dark-colored honey that has molasses-like aromas. The honey produced by the Linden tree flowers has a minty taste and light color. When honey is collected, a small amount resin can end up in the honey. To seal the elements out of the hive, bees use tree trunks to drill resin. Propolis is formed when tree resin comes into contact with bees' mouths. The seal of propolis is broken when a hive's lid is opened. This can make the honey taste mildly nutty or piney.

BEEKEEPING EQUIPMENT

Investment in materials is a must for any hobby. Beekeeping is no different. You will need to invest in some equipment. The equipment you need will depend on the size of your beekeeping operation, the number of colonies and the type of honey you intend to produce. You will need the basic equipments: a hive, protective gear and smoker. The hive device components are also required. The hive, a man-made structure in which honey bee colonies live, is the hive. Over the years, a wide variety of hives has been made. The Langstroth, or traditional ten-frame hive, is the most popular choice for beekeepers. The standard hive is made up of a hive stand and a bottom board with an entry cleat, reducer, and a set of boxes, or hive bodies, with suspended frames that contain honey or base. Both the interior and exterior covers are provided with measurements to allow for individual hives. You can separate the hive bodies that contain the brood nest by removing the queen from the honey supers (where surplus honey is kept).

The hive stand is an optional piece or hardware that raises the hive's floor (or bottom board) from the ground. This reduces humidity, improves bottom board life, and keeps the front door free of grass and other weeds. Concrete blocks, railroad ties and bricks are all possible options for a hive stand. A hive stand may support one colony, two or more colonies or a multi-colonial network.

Bottom board - This board is for the colony's ground and serves as a landing and takeoff surface for bees. To prevent rainwater from entering the hive, the colony will be slightly reversible so that the bottom board is exposed at the front. Many bee supply retailers sell reversible bottom boards. They have a 7/8-inch or 3/8-inch front gap.

Hive bodies – The standard hive body comes with ten frames and is available in four depths. Most commonly used for rearing larvae is the full-depth hive, which measures 9 5/8 inches in length. These massive units have ample space to support large brood fields with minimal interference. They are also ideal for honey supers. They can weigh over 60 pounds but are difficult to hold if they're loaded with sugar.

The super medium-depth (also known as super Dadant or Illinois) is 6 5/8 inches high. This scale is ideal for honey supers. However, standard-sized timber cannot easily be cut. A few beekeepers prefer an intermediate size (7 1/2) "between full- and medium depth supers," especially those who make their own boxes.

Ultra-shallow depth, at 5 1/6 inches tall, is the most lightweight unit (about 35 lbs when filled with honey). This size is the most expensive for assembling usable space comb space per square inch.

The section comb honey is 4 5/8 inches tall and can be used to preserve either section holders, plastic rings, or basswood section boxes. Section comb honey processing is an extremely skilled art that requires extensive management. It is not recommended for beginners.

Hive bodies with eight doors are a common choice for beekeepers. Although this was mostly imported, one U.S. manufacturer now sells eight-frame boxes as hive box boxes for the English greenhouse. For raising queens and selling small starter colonies (nucs), beekeepers prefer to use three to five nucs, which are usually standard-sized frames. These can be purchased from bee manufacturers. They are made of wood or cardboard and are only temporary. Different management strategies can be used depending on the depth of the individual hive bodies. One strategy is to have one full-depth hive, which could give the queen the space she needs to lay her eggs. For food storage and optimal expansion of the brood, additional space is necessary. A single, full-depth chamber for brood is usually used by beekeepers when they choose to crowd bees in order to produce comb honey. This is also the case when a kit is constructed or when a nucleus colony is created. Most beekeepers prefer two full-depth or one shallow hive body for the brood area. The exchange of combs is possible between identically shaped bodies by using two hive body shapes. If you are looking to escape large, deep hive body beekeepers can use a brood nest that has three shallow hive parts. This is a good option, but it's also the most time-consuming and expensive approach to assembly. It requires thirty frames and three boxes.

Frame and combs-The frame holds the beeswaxcomb suspended in a frame. This is the fundamental structural component of the hive. A sheet of beeswax, or a plastic base, is used to create the wooden or plastic beeswaxcombs in human-made beehives. After workers have removed wax from the base, the cells are drawn and used to store honey and pollen. To match the different depths of the bee-body, the frames are 17 5/8 inches in width and can be 9 1/8, 6 1/4 or 7 1/4 inches thick. A frame is made up of a top, two end bars and a lower bar. The top bars can be grooved or wedged. Bottom bars can be broken, firm, grooved, or grooved. While some forms have their advantages, others may be more cost-effective. The top bars at the ends are suspended from rabbets or ledges. For reinforcement of the recess, spacers or metal strips in V-shaped shapes are often nailed. Common commercial end bars have shoulders to ensure that there is enough space between the adjacent frames and the box hand.

The comb base is made up of thin sheets beeswax with patterns of worker-sized cells imprinted on one side. The comb bases have different thicknesses. Thinner foundations can be used to make chunk honey, segment comb honey or cut-comb honey. A heavier, more solid base will be used in frames and the brood chamber to extract honey. Thicker foundations can be used to reinforce vertically embedded wires, metal edges and thin sheets of plastic or nylon threads. When deciding whether to invest in plastic foundations in plastic frames or pure wax foundations in wooden or plastic frames, there are many factors you need to consider, including the initial cost, installation time and longevity as well as the planned use duration. These plastic foundations and frames are more popular. Protect the foundation within the frame using either horizontal wires or metal support pins. To protect the base from being damaged by wire hooks, the top bar has a thin wedge. You can strengthen your combs by inserting horizontal wires (28 and 30 gauges) through the base with an electric current. It is difficult and time-consuming to do this because even a well-supported foundation contributes to well-drawn hairs. A new foundation should not be used to frame colonies that are rapidly growing, such as a swarm or package or split colony (division), or established colonies. By applying wax to the foundation sheet imprints on six-sided cells, workers create beeswaxcombs.

Four pairs of glands are located on the undersides of worker's abdomen that form beeswax. The wax is then pressed into wax scales after it has been released to the sun. The bees remove the wax scales from their abdomens and place spines on the middle legs to create a comb. The wax scale is transmitted to the mouthpieces, where it is processed until it can be folded into six-sided cells.

Queen excluder – The queen's main role is to confine and rear the queen, as well as to store pollen for the brood nest. This valuable piece of equipment is rarely used by more than 50% of beekeepers. The queen is often referred to as "honey exclusionrs" by beekeepers because workers sometimes find it difficult to get through the small openings in the excluder to store honey in their brood nest. To reduce this problem, encourage beekeepers to store nectar in the supers prior to mounting the excluder. The excluder will be attracted to nectar that is placed in a drawn-comb. Supers of foundation should never be placed above a queen excluder.

An excluder is a thin layer made of perforated plastic or metal with enough holes to allow workers to pass through. Many designs have welded round-wire grills that are backed by metal frames or wooden frames. The honey frames, located in the super-abundantly above the brood chambers or comb sections, serve as natural shields to protect the queen. The queen exclusion function would be the same if the brood chambers are reversed in spring timing is correct. Queen excluders are used with the first supers attached. However, the supers should be mounted only after any nectar has been deposited. A queen excluder can help prevent honey from darkening because the beeswax used to make brood darkens.

In a two-queen scheme, queen excluders can be used to separate queens, collect queenright colonies and prevent emergency swarming. A queen excluder can also help you find the queen. After 3 days, you can determine which hive contains the queen by placing an excluder between them.

Inner cover - This inner cover is located over the outer telescopic covering of the super. It prevents bees using wax and propolis to nail on the outer super cover. It provides isolation airspace just below the outer cover. The inner cover shields the interior of the hive from direct sunlight during summer. This protects the moisture-laden air from coming in direct contact with cold surfaces during winter. To help bees extract full honey supers, a Porter bee escape can also be added to the middle hole of the inner cover.

Exterior cover - A telescopic exterior cover protects the hive parts from adverse environmental conditions. This covers the interior cover and the top edge the uppermost hive bodies. To prevent leaks and weather damage, most roofs are covered with a metal layer. The outer cover can be removed, but the inner cover must be left in place. This reduces the disturbances to the hive, and allows the beekeepers to smoke the honeybees before handling the colony.

Clear covers are used by beekeepers to cover the hives when they are being moved frequently. Also known as a "migratory cap", The cover is flush with the sides and can be stretched over the ends. These covers are lightweight and can be removed easily. They also allow for the stacking of colonies. To secure a load on a truck, it is important to stack the colonies tightly.

Additional equipment - You can add other pieces to your basic hive components. Some beekeepers prefer the slatted base plate. Others are painted with an English pattern. Beekeeping offers a lot of freedom for creativity and individualization.

Painting the hive parts - Paint should be applied to all areas that are exposed to the elements. The hive should not be painted inside. Instead, the bees will varnish it with propolis, a mixture of wax and plant sap. The sole purpose of painting is to preserve the wood. Exterior beekeepers often use strong, white latex paint. It is best to use a light color in summer as it prevents heat buildup in the beehive. White is a common color but different colors can be used to help reduce colonial drift.

Tools made from plastic - The main parts of the hive were traditionally made of oak, redwood, or cypress. Today, all hive components are made of plastic. Plastic hives and frames that snap together are strong, sturdy, lightweight, simple to assemble and require very little maintenance. Plastic frames and floors are becoming more common. However, plastic hive covers, bottom boards and hive bodies have not been as effective because plastic doesn't breathe and does not allow for quick ventilation. Plastic can easily warp and some plastic forms let too much sunlight in, making it difficult to draw base.

Equipment suppliers - Most new bee equipment comes "knocked down" (or unassembled) when it is purchased. However, you can purchase assembled equipment at a higher cost and with a longer shipping fee. The bee supply dealer will usually provide instructions and a manual for installation. It is a good idea to get help from a more experienced beekeeper before installing the components of your hive. Beginers should purchase their equipment as soon as possible to be able to put together hives and paint them before the bees arrive. The foundation sheets should not be placed in frames until they are ready to use. High temperatures and storage can cause wax to warp and spread, which can lead to poorly drawn combs. Many beekeepers believe that making or purchasing used tools can help them save money. All methods will require the appliances to be used on a regular basis. When designing beekeeping equipment, it is important to have a good understanding of the bee-space. As a guide, you can review existing building plans and use industrial parts. While most beekeepers believe they are capable of making roofs, hive heads and bottom boards quickly, frames can be more difficult and time-consuming. The quality and cost of the supplies, as well as the tools and woodworking skills of the beekeeper are key factors in determining success.

It is not a good idea to buy used equipment. It can be difficult to find the right equipment, or evaluate its value and importance. Secondhand equipment can also be contaminated by pathogens or may have been stored for a long time. This can cause various bee diseases. A certificate of inspection is often requested to show that the state inspector has examined the hives and found no evidence of disease.

Chapter 10: Important Beekeeping Guidelines

Even if you are a beginner, there are many things you need to remember when beekeeping. These are the basics.

How to Respond When Bees Arrive

What do you do if the bees arrive at your door?

Breathe. Then, follow these instructions:

You should inspect the package carefully. You should check whether the bees inside are still alive. Sometimes transportation can cause the bottom ones to die. If this happens, contact your vendor so that they can replace the bees.

Avoid putting bees in your car's trunk. The shade could suffocate them and cause them to become thirsty and agitated.

Once the package arrives at your home, spray cool water onto it. Spray the package with a spray bottle.

After spraying water on the package, let it cool for at least one hour.

After an hour, spray the package with non-medicated syrup. Then, move the bees into the hive.

Beekeeping

You might also be curious about what the bees eat. You can remember that bees eat nectar and pollen, so you need to ensure you have the right flowers and plants around.

Because pollen is one of the most nutritious natural foods on Earth, it is essential. It contains all the nutrients that honeybees need to survive. Nectar, which is sweet and tastes similar to honey, is what bees love.

Honeybee larvae can also eat honey from the hive. The queen may decide to give them Royal Jelly sometimes, which is packed with more nutrients. It contains fertility stimulants, which help young worker bees grow. The queen can also eat it, so she will be able to have healthier babies in the future. Royal Jelly can help a queen lay as many as 2,000 eggs per year.

Where should they stay?

It is not possible to just place the beehive where you want it without considering the implications for the bees.

Remember that bees seek balance. They are not happy to be in places that are too hot or cold. You need to ensure that they get enough sunlight but not too much, and that they have the right amount cool air.

You shouldn't put up a beehive if you live on a hilltop. It might windy and not be suitable for them. Avoid extremely low areas as they may not have enough cold air and flooding could also be an issue.

Important is also the availability of food sources. Some colonies have syrup they can drink. However, it is important to keep them close to water sources and plants that contain nectar. These attract bees and lead to proper pollination. You'll find out exactly which plants and flowers you require in the next chapter.

How many hives are there?

It's fine to start with one hive as a beginner. Once you are able to let bees produce honey, wax, and pollinate, then you can decide to buy a second one.

How to Maintain Hives in Different Seasons

Also, keep in mind that the beehive must be maintained for all seasons. These are just a few suggestions.

Autumn

Queens are more likely to leave the hive in fall so be sure to check inside. Good tip: Look for eggs. Eggs, not larvae, are the best option. The larvae are more than two days old, which means that the queen may have already left.

During this time, it is important to feed and medicate. Remember to verify that there is enough honey. To keep the bees safe from infections and diseases, you can also give Teramycin or Teramycin as Antibiotics. The bees prefer 2-to-1 sugar syrup.

To prevent freezing point, wrap the hive with tar paper, especially if you live near extreme cold. You can also stop too much sun from entering the hive. This season, it is important to regulate the temperature.

A good ventilation system is essential. You need to maintain a temperature between 90 and 93F so that the warm air from the cluster doesn't rise, but the inner cover doesn't get cold.

As mice are very common in this season, you can add a mouse guard at the entrance to your hive.

You can put on a windbreaker if the weather is extreme. If you want, you can also use burlap or fence posts.

Winter

It's a good idea to inspect the entrance of your hive during winter. Check that there are no dead or snow bees blocking the entrance.

Give the bees plenty of food. They find it difficult to leave the hive at this time of year. You can check their food supply by looking inside the hive. You don't have to disturb them; just give them a peek.

If you don't see honey in the top frames it is a sign that they are not getting enough food. This indicates that they need emergency feeding.

It's best to buy packaged bees if you plan to begin taking care of your bees this season.

This is also a great time to make beeswax candles and brew mead.

Make sure that you have enough equipment for the winter season and that you are ready to go.

Sommer

In summer, you may need to visit the hive at most every other week to assess the health of the queen and the rest of the bees.

As needed, add more honey supers.

Keep an eye out for any other insects that might be stealing honey from the bees and make sure you keep them away. You must always be vigilant and not wait for the hive be completely infested before doing anything.

Swarm control is important up until mid-summer. You can relax in the late summer, when swarming is not as common.

Make sure you harvest your honey crop before the end of the nectar flow. You will need 60 pounds of honey. Now, it is time to get your gloves on, as the bees are extremely protective of this honey.

It's also a simple season because you don't have to wait until the end to get all the honey.

Spring

As soon as possible, inspect the colony in the spring. It could also be a sign that the bees have become too cold to inspect.

You might be able to see or hear the clusters of bees. Some may not have survived winter. It might be bad news if you can't hear any buzzing sounds.

You should check if they are eating. It is important to check if the bees can eat properly. You should cap honey with white capping. If you don't see any, it's time to feed emergency food.

Take a sirop and medicate during this season.

Be aware that bee colonies might flourish during spring season if the queen is able to reproduce many. This is a possibility. You should add queen excluders and honey supers to this period, otherwise the worker bees may not make it through.

You should look out for signs of swarming. More information on this later.

Price range

It is important to know how much money you can afford to invest in this type of business.

It's not expensive and you can breathe a sigh relief. It is common for hives to cost between $200 and $400. However, it all depends on the type of bees you use. A package of bees will cost you between $60 and $80. These are your one-time expenses.

Imagine that each beehive produces 60-90 pounds of honey per year. This would mean that you could get anywhere from $5 to $7 per pound of honey or $600 to $700 per beehive.

Zoning and legal restrictions

It's a good idea to consult your local ordinances before you decide whether or not it is okay to keep bees at your home. Some laws allow only one beehive. Others prohibit you from having more.

You should also check if your hives could be registered or if this is not necessary.

Think about the environment

If your local ordinances allow it, you can keep bees almost anywhere. However, be aware that you are not allowed to keep bees in areas like the desert or near tundra. You get an extra bonus if you are located in tropical regions.

Space requirements are not very important, although you should refer to Chapter 3 for the size of your hive. Proper spacing is essential. These are two completely different things.

Don't worry about it, bees can return to their hives even if they travel 6000 miles to pollinate. It is in their nature.

Give them time and effort

Another important thing to remember about beekeeping is: It is important to be patient with the bees.

The colony does not need to be your primary focus. It is important that it receives some time, but 35-40 hours per year is still a small amount compared to the time you spend with other animals.

Installing the hives takes very little time. You don't need to spend your entire weekend on it. You could visit it 7 to 8 times a year.

You don't need to be told how to look at your honeybees, especially if you already have them with you. It will come easily. It's important to know that you are responsible enough and will not let your hard work go to waste.

Chapter 11: The Negative Effects of Chemical Insecticides on Bees and Their Honey

Chemical insecticides pose a grave threat to bees. They wipe out colonies and reduce bee procreation. This can also negatively impact agricultural products.

The chemical insecticide, e.g. Gamalin 20 kills bees, and makes honey unfit for human consumption. Chemicals used in honey harvesting can cause gastro-intestinal diseases and other health problems.

For easy harvesting, some farmers use chemical insecticides while others set fire to the hive to prevent bee stings. However, this is without considering the potential negative impacts on the society. A farmer who does not use the recommended protective clothing for harvesting can only be stung by bees.

DEFENSE and COMMUNICATION

Defense: Honeybees live in colonies, where workers sting intruders to defend them. Alarmed bees also release a pheromone which stimulates other bees' attack response. Honeybees can be distinguished from other Hymenoptera species by small barbs on their stings. However, these barbs are only found in worker bees. The sting apparatus and its associated venom sac are modified to allow the bees to pull out of the body once it has been lodged (autonomy). Once detached, the sting apparatus also has its own musculature, ganglion and muscle, which allows it to continue delivering venom. After the sting is lodged, the worker dies and the venom is torn from the abdomen. Honeybee venom is known as Apitoxin. It contains several active components. The most powerful of these are melittin and the most dangerous phospholipase. Complex apparatus including the stinger's barbs is thought to have been created in response to vertebrates predating on honeybees. The sting apparatus doesn't usually detach and the barbs don't normally function unless embedded in fleshy tissues. The sting can penetrate the membranes between the joints of exoskeletons of other insects and is used in fights among queens. However, Apisceranajaponica provides defense against predatory wasps such as Apisceranajaponica. The Asian giant hornet is normally protected by a large number of worker bees. They vibrate vigorously to raise the temperature and kill the intruder. Intruding wasps were previously thought to be killed by heat. However, recent experiments have shown that the increased carbon dioxide levels in the ball caused the deadly effect. This phenomenon can also be used to kill queens that are perceived as being intruding on or deficient. It is known as "balling the queen" by beekeepers, after the formation of a ball of bees. For honeybee species that have open combs (e.g. A. dorsata): Potential predators are warned by a "Mexican Wave" that appears as a wave across the surface of a layer bees packed in the comb. It is composed of bees arching their bodies and flicking the wings.

COMMUNICATION: Honeybees communicate using a variety of chemicals and odours. However, they also use specific behaviours to convey information about the environmental quality and location of these resources. The signalling used varies from species to species. For example, Apisandreniformis honeybees dance on the top surface of their comb. This is horizontal, not vertical as in other species. Worker bees then orient their dance in the direction of the resource they are recruiting. Apismelliferacarnica honeybees prefer to use their right antenna for social interactions.

METHOD OF TRAITING BEE STING

You can remove the bee sting apparatus by using your finger to rub the affected areas. As other bees can sense the odour, they may continue to sting the animal if the bee sting organ (poisonous) isn't removed. An animal's sting organ can become swollen for up to two days before it dies. If not taken care of immediately, continuous stinging by bees can cause coma or even death. Bees are not capable of stinging higher animals such as humans, monkeys, Chimpanzees, Gorillas and elephants.

You can still treat bee stings with orange lime (Oromaikirishy). Simply cut an orange lime and rub the juice onto the affected area. For approximately two to three days, repeat the process. If swelling persists, consult a doctor.

HONEY IS IMPORTANT

HONEY: This complex substance is made from the nectar and sweet deposits of plants and trees. Honeybees collect, modify and store it in honeycombs to provide food for their colony. Apis species have all had their honey collected by native people. However, for commercial purposes only A. cerana and A. mellifera have been used to humans from the nests at various gate spoilt. You can use honey to reduce hunger.

It aids in brain development

Clears the throat

It is used to treat cough.

It can also be substituted for sugar in food preparation.

It is also useful in the pharmaceutical industry.

It can also be used to infuse pain during child birth.

It can be used to heal wounds and inflammations.

It can be used to treat burns.

It is used for hair treatment by women.

It is used for clearing dandruff.

It can also be used in the manufacture of cream, soap, and other industrial products.

It is used in bakery production.

It can be used to treat all types of bacterial infections.

You can use bee wax to make candle wax or polish.

For the manufacture of shoe soles, you can also use bee wax.

You can bait (or lure) hives with honey bees.

It can also be used to make cosmetics.

It can also be used to treat fungal, viral and other diseases.

Chapter 12: What Beekeepers Need to Know

Beekeeping and the Law

Although beekeeping can be done on a small scale in a backyard setting, it is important to first check if there are any local ordinances or zoning laws.

You should ensure that your Homeowner's Association does not prohibit beekeeping if you are a member.

It is rare for beekeeping to be illegal in the United States. Some areas, however, have laws that place practical restrictions on beekeeping. There are restrictions on the number and type of bees that can be kept, as well as requirements that beekeepers provide water for them.

Most states have "nuisance legislation" which is intended to make illegal certain activities, such as barking dogs or strong smells.

Use common sense to think about your neighbors, and to not allow beekeeping to have an impact on their lives. It's a great idea to share a jar of honey. You should be aware that bees can cause fear in some people, particularly if they have an allergy to bee stings.

To prove that your hives are not a threat to the environment, be prepared to provide educational materials. There will always be bees around our towns and cities if there is nectar- and pollen-producing plants. If beekeeping was illegal, you'd see more wild bees and the bees wouldn't disappear.

Also, it is important to speak with your agent about insurance. Ask about adding coverage for accidents. Additional insurance coverage could be very beneficial if you are held responsible for these incidents.

If you have any objections to beekeeping, or if there is zoning law that prohibits you from doing so, it may be possible to relocate your bees and beehives to another location.

Many beekeepers are unable to keep bees at their home so they have taken the initiative to keep them on local farms.

Local beekeeping groups may have information that can help you find a suitable location to keep your bees. You might also contact local fruit and vegetable growers and gardening clubs to find a suitable spot.

Every gardener will understand the importance of pollination.

What is the Time It Takes to Beekeeping?

Be sure you're ready to commit to beekeeping before you do.

The amount of time you spend caring for your bees can vary from one year to the next. In late spring and summer, the average time spent on your first colony will be about an hour per week. Each additional colony will take about half an hour. It may take longer some weeks than others.

You will spend your time taking off protective clothing, lighting a fire, and assembling the necessary kit. It doesn't matter if you have one or many colonies, this takes time. Additional requirements include feeding, bottling and extracting honey.

There is not much to do during the later Fall, Winter, and early Spring. This is the time most beekeepers use to prepare for next season.

Experienced beekeepers will find it takes less time. It is possible to expect to be faster over.

Time.

Names of honey bees scientifically

There are four main species of honeybees: Apis Florea, Apis Dorsata, Apis Cerana, Apis Cerana and Apis Mellifera. The Apis Mellifera honey bees are found in North America.

Apis, Latin for "bee", is the genus name. Mellifera means honey producing. This species tends to produce large quantities of honey that can be stored over the winter months. Apis Mellifera is essentially honey-producing and pollen collection.

Why do Bees Sting?

There are many uses for bees' stinging abilities, depending on their role. The drones (male bees), are not able to sting because they don't have stingers. The stinger, which is a modified egg-laying device (egg ovipositor), is only available to females.

Workers bees use stinging to defend their colonies. They will only resort to stinging if they feel threatened.

Only one species of bee can die from stings: the honeybee. The barbed stinger cannot be pulled out of a person by a honeybee that has stung them. It leaves behind the stinger as well as its stomach and digestive tract. This is an abdominal rupture large enough to kill the honeybee.

Queen bees are different from worker bees in that they sting for very different reasons. They sting to eradicate competition in the colony. The worker bees will raise new queens when a colony becomes "queen-less". There will be several young queens at once. They will fight until one queen bee remains. They use their stinger against the queens.

It is rare, however, for a queen to sting someone.

How honey bees communicate

The communication between honey bees can be fascinating and essential for beekeepers to understand.

Two main methods are used to aid communication. The first is chemical and the second is choreographic.

Honey Bee Pheromones

Animals produce chemical smells that trigger behavioral responses in other species. These chemical scents are known as pheromones.

Honey bee pheromones are essential for the survival of the colony. Each type of bee produces a variety of pheromones that trigger certain behaviors. Examples include;

Queen pheromones, also known as queen substance, tell the colony when the queen is at home. This activates many worker bee activities.

-Excluded from the hive, the queen pheromones are used to attract mates and influence the behavior of male drones.

-Beekeepers use pheromones at the entrance of the hive to guide the foraging bees to their hive. This attractive scent is due to the 'Nassanoff gland' located at the tip the worker bees abdomen.

Worker bees use alarm pheromones to warn them of aggression towards their colony.

-Pheromones are produced by developing bee larvae and pupae that enable worker bees recognize the gender, stage of development, and food requirements of their brood.

Dance of the Honey Bee

The honey bee's unique dance is the most fascinating language to humans. Foraging worker bees perform a variety of dances to communicate their messages.

They bring back news of nectar, pollen or water to the hive and must share this information with fellow worker bees.

The dance style provides surprising information about the food the foraging bees have found.

There are two types of dances to be aware of: the "Round Dance" and the "Waggle Dance".

The "Round Dance" tells worker bees there is food nearby (at least 10 to 80 yards).

The "Waggle Dance" communicates that the food source is located further away from the hive. The dance is a side-to-side motion of the abdomen while the dancing bee forms an eight-pointed figure. The strength of the waggle and the number of repetitions, as well as the direction of dance and sound made by the bee together, communicate incredible details about the source of the food.

During dances, bees stop to offer a taste of the food that they brought back. This gives further information about the origin of the food and the time it was made.

What do Honey Bees Eat?

Honey bees eat pollen and nectar from many flowers, which allows them to produce honey.

The result of various flowering plants is pollen, which is a fine powdery and dusty substance. Pollen is a rich, natural food. It contains all the nutrients that a honeybee needs, including sugar, carbohydrates and protein. The honey bees take nectar from flowers to make sweet syrup. Honey bees take nectar from flowers and turn it into honey.

Honey is the preferred food for honey bee larvae. Royal jelly will be given to larvae who are selected to become future queens.

Royal jelly is a white liquid that is secreted by young worker bees' mouth glands. It is made up of pollen and chemicals from worker bees' glands. It is very rich in protein and fatty acid.

Royal jelly is given to workers and drones during larval development. Future queen larvae eat the royal jelly throughout development.

The queen honeybee is fed royal jelly and grows twice as fast as an ordinary honeybee. Queens can live for as long as three to five years due to royal jelly's high nutritional value. This is a significant improvement on the life expectancy of worker and drone bees. Queens can also lay as many as 2,000 eggs per day.

The Honey Bee's Lifespan

The expected lifespan of the honeybee depends on the species of bee.

Queen bees can live for up to three decades.

Drones live a shorter life span - either they die when they mate or they are expelled from their colony, where they will soon die.

The time of the year is a major factor in determining the worker bee's life expectancy.

A worker bee will only live for 6 weeks during the summer, spring and fall. The worker bee's primary function is to care for the larvae and pupa for the first two weeks. After that, they will forage for food and collect pollen until the time they die.

Workers bees born in late fall can live up to 4 - 5 more months. They have another purpose, which is to care for the queen and keep warm in the winter.

Chapter 13: Separating Fact from Myth

As we have said, beekeeping is not for everyone. However, some of the information they will give you about apiculture are based on myths that were based upon fanciful claims about the world.

Nature is kind

This is right up there alongside "life is fair".

You should also know that not all honey is equal in terms of providing adequate nutrition for your colonies to survive winter. It can be dangerous to be too focused on "natural" or organic methods.

It's easy to survive

Look at the poor dinosaurs!

A colony in danger can pose a threat to its health, both in a cultured and wild population.

Even in natural environments, bee colonies don't grow in size.

The hive structure is the key to success

The best hive for your colony is the one that you choose.

A beekeeper should not have ever had to deal with the collapse of a colony.

All chemicals are bad

Each person who reads this has at one time or another taken a medication to treat an infection or a headache.

The truth is that bees actually dig for certain chemicals produced by nature (yes, even nature makes chemicals!). These chemicals are used to control their hives and to eliminate health risks.

But there are different levels of chemicals.Some of these really are bad.This is especially true of the chemicals found in pesticides and other androgenic pollutants.Certain miticides (take care) will also contain some of these harmful chemicals, but they're not recommended here.Naturally-derived chemical treatments, like the ones listed here in the mite management section, serve the same purpose as many of the natural treatments human beings have used for millennia to combat infection and disease.And let's face it, the invention of penicillin has saved the lives of billions of human beings.

Chapter 14: Opening the Hive

Opening a beehive can be a very enjoyable experience for any beekeeper, novice or expert. While opening the hive can help the beekeeper relax from the stress of daily life, it can also have an adverse effect on the bees. The honey bees feel anxious when they are interrupted from their home. They need to recover for hours, if not days. This is not good for honey bee productivity and wellbeing. Imagine someone going to your house and looking through everything. They would then take your belongings with them and leave you to tidy up the mess. After you've settled in and things are back to normal, they will return week after week for another inspection. This increases everyone's anxiety. Once we've decided that it is important for the beehive to be opened for inspection, we can concentrate on what we will find. There are high chances that we will get distracted by what we see, and miss what we want. It's normal to lose your sense of direction. It's like talking on the phone while driving. There are many reasons for opening the hive. This can cause some to be inconsistent, so only a small portion of the assessments should be recorded. How often should the hive be opened by bees? It is only necessary and important to open the Beehive every so often. The Beehive should only be opened in cool weather, which means not too cool, not too hot, and not too breezy. When you'd be comfortable sitting outside in shorts and a T shirt. When it is breezy, raining, or with an approaching electrical storm, try to keep the hive closed. Breezy weather can cause brood to become shivering and it is best to not lift brood frames.

The beehive should only be opened at certain times, such as:

IN THE SWARMING SEASON, the hive should always be open for inspection every seven to ten day, from September to December, to determine if preventive steps should be taken to stop swarming. If the honey bees have started queen cells, this gives the beekeeper enough time to take action.

DURING THE REST SEASON - January through March/April depending on the success of scavenging movements, a two- to four-week inspection should suffice to ensure that everything is in order. If you have a steady honey stream, you might need to build it up and remove some honey to increase your stockpiling.

DURING OFF-SEASON - From April/May to August, as long as the honeybees have sufficient sustenance storage for winter, they should not be disturbed. It is impossible to open the hive during the cool season or inspect frames for food. To check the weight of your hive, simply lift it towards one side. A single box 8-outline honey hive should contain at least five honey frames. The normal net weight for each honey casing is 2.4 kg. This adds up to 12 kg. A single box hive should weigh between 24-30kg to be winter-prepared.

WHAT TO DO DURING INSPECTIONS

You can do a lot of routine maintenance and checkups on a beehive without having to remove every frame. Simply lift the top to add nourishment, dust patties or bug medications. To check for swarm cells, you can tip a broodbox up and examine the base. By lifting the back of a honey case, you can check the weight. A sticky board can be used to check for insects. A sticky board can be used to check for bugs in a hive.

These checks should be performed as a routine when a hive opens for inspection. These routine checks may change the goal of the next Frame Inspection. If you were originally going to inspect the queen and her laying performance, but now you notice foulbrood, it is time to inspect the hive.

First, check your nose. You should wear a cloak. Lift the top and place your nose over the hive to see if there is any air rising from the inside. If you detect an unpleasant or foul-smelling odor, such as Sauerkraut, this is a sign that your hive is infected with European Foulbrood, American Foulbrood, or both.

While you are removing the cover, observe the reaction of the honeybees to the frame and mat. If you find that a large number of honeybees are welcoming you by raising their stomachs, exposing the

Nasonov organ, and exchanging excited fanning movements with their wings, it is best not to open the hive again. Honey bees will let you know that they are having a bad attitude.

Before putting the top on the ground, make sure you check that the honey bee queen isn't under the cover. There is a possibility that the queen or a few honeybees may try to escape from the passageway if you blow smoke into it.

After you have removed the cover and hive mat, gently blow smoke from the top of the frames. You will hear the honey bees buzzing, which is usually a quiet buzz. Get used to that sound. If the buzz becomes louder, noisier, and more intense than the quiet buzz, then it's likely that the province has lost its queen. It may take many years before a hobby beekeeper encounters the situation of a queenless colony. This can be confirmed by looking for the queen, or evidence of her presence. Eggs and young larvae.

If you see combs under the cover, this is a sign that honey bees need more space to make combs. No matter what your reason for inspecting the hives, you can add an honest activity mindset to allow them more space. Add honey frames to the establishment, or stack a super-on with frames and establishment.

These checks are performed simultaneously while inspecting the frames. They are not recorded. The typical procedure for checking frames is to start with the second outline, which uncovers the honey stores.

Are the combs made of honey or dust? The outside combs usually contain honey while the inside combs have honey at the top corners. For combs that have brood, a few cells of dust filled with dust fill the space between honey and brood. Honey bees might require sugar syrup if there is not enough honey.

Is there brood during all stages of improvement? The size of the brood patches can vary from very little to none in the outside combs to large areas in the middle combs. They often fill almost the entire edge. In all stages of improvement, such as. There are eggs, larvaes, young honey bees and topped cells. A lot of eggs and young brood is a sign that the queen does a good job and there is no reason not to see her.

Is the brood top normal? Or is it slightly curved? Healthy, well-topped brood cells with a healthy top are slightly raised. This is a sign that they are full. In some cases, the cappings will be indented in at the focal point. This indicates that brood is in poor condition. To recognize brood illness, you will need to look for other signs.

Are some worker cells topped with drone cells? If there are a few topped specialist cells that stand out above the rest, and the topped cells look like projectiles, then it is possible the queen is old or debilitated, and is laying unfertilized egg into worker cells. The drone cells are not to be confused with the usual ones, which look almost like shots but are 2mm larger in width. Sporadically, mated queens have euthanized a few infertile or fruitless eggs.

Do honey bees store a lot of honey in their brood? If there is any, the brood box usually contains a large patch of brood and honey in its top corners. If the brood zones are separated by honey patches, honey bees might need more space. You can either add honey frames with establishment, or stack a super-on with frames and establishment.

Is there no egg or young brood? If there aren't any eggs or young brood, it means that there is no laying Queen. There are two options.

1. If there are queen cells available, the state might plan to swarm the queen and stop bolstering her to thin her down so she can fly with them. Or the queen was killed accidentally or lost when you last opened the hive. If you have spent some time looking at the hive and found no new brood, it is necessary to request a queen immediately. The packaging is made by placing honey bees down a large channel and then adding a queen that is not the mother to any of them. The supplier may have any stock, but the honey bees will come from it. It doesn't really matter that they are there to care for the Queen and

her children, since their lives are over. The Queen should be mated to a specific race of drones, however. The honey bees have a new queen.

2. If you find queen cells that are open and have the top cut off, it is likely there is another queen in the hive. She will lay eggs in about a week.

3. If your hive does not have queen cells, it is likely that there is no one to help them. If you are unsure about the situation, you can embed a comb with eggs of another hive and then check to see if the honeybees have started raising queen cells.

Also, you should check if queen cells are present. If you don't make preventive moves, you might lose a swarm during swarming season. Queen cells can happen at different times. This could be due to the queen dying or being lost accidentally. Or the honey bees might have decided that the queen is too old and are trying to find a replacement.

Is there a gap in one of the brood caps? Punctured cell tops or cell tops that have been completely opened or evacuated are a side effect of all honeybee brood problems. This indicates that something is not right with the topped brood. Some productions show this as AFB. However, there is no need to panic yet. This indication is common to all events of dead brood in topped cells, even those that have not been afflicted. The honey bees open a portion the topped cells for examination when normal development of honeybees from the pupae has ceased.

Are there discolored or stained larvae? Healthy larvae will appear silvery-white and flickering from the outside. If the larvae are discolored, it is likely that the brood has contracted a disease.

If you find something amiss in your honey bee hive, you must report it to the Department of Economic Development, Jobs Transport and Resources (DEDJTR). It is easy to spot a few health issues in your honey bees, such as AFB. It can be difficult to identify a honeybee illness.

After looking at your beehives for some time, you may still be unsure of the problem. Ask a more experienced beekeeper to help. Be sure to determine the cause of the problem using all available methods. Do not keep your fingers crossed and hope it will go away.

Chapter 15: Parasites & Pests

There are many types of parasites and pests that can affect your honey bees. You can learn about them to be the best beekeeper you can.

Mites

The mite is a common parasite. They feed on the body fluids of the larvae and adults. You can see them with your naked eye. They are small, brownish or red spots on the throat. These mites can also be carriers of viruses that can cause serious harm to bee colonies. These viruses can cause wings deformities in bees when they are infected.

Although these mites are more likely to concentrate on drones than the honeybees, it is not always true. They can cause bees to become weaker and reduce their lifespan. The wings will make it difficult for new bees to survive. Protecting your bees from mites is a must as they can spread quickly from one colony to the next.

Mite Treatment

There are many ways to treat mites that are affecting honey bees. Apistan and CheckMite are two options if you prefer to use chemicals. According to the EPA, these chemicals can be used correctly and can kill a large number of mites. However, these chemicals are not likely to cause any harm to bees and should not be used to alter their behavior or reduce their lifespan.

You can also use a mechanical approach to control mites, which interrupts their life cycle. Although it won't eliminate all mites, it can reduce the likelihood of an infestation. Varroa Mite Control Entry (VMCE) is a device that is frequently recommended. It attaches to the bottom of the hive with a screen and will remove mites when the bees enter or leave the hive.

Detection and Treatment of Mites

You should be on the lookout for red or brown spots on your bees. These could indicate mites. You can also identify mites by the changes in the formation of the bees' wings. This type of infestation spreads quickly so it is important to act immediately to get rid of it. Because mites can easily spread from one hive of bees to another, it is important to treat all colonies.

Nosema

Nosema, a microsporidian, is one type of problem you should be looking out for. It can cause intestinal problems in adult bees. This is often associated with Black

Queen Cell Viral. This problem is usually caused by bees not being able to move out of the hive to eliminate any waste. This can happen during cold spells or winter months when the bees have been kept inside the hives.

Because it spreads very quickly among bees, nosema can prove to be dangerous. An epidemic can occur in as little as a few days. It is because the spores of an affected bee are then consumed by another bee, and it is difficult to stop the chain reaction.

Nosema symptoms

Nosema has a few major problems. There aren't any obvious symptoms that a beekeeper can see. A beekeeper won't realize that they have a problem until the effects of the pathogen have been seen and the bees have died from them. If the cold weather persists, be prepared for it.

The main cause is that the bees are unable to leave the hive to eliminate waste.

Nosema Treatment

To successfully treat the Nosema hive, you need to increase its ventilation. It is also necessary to use antibiotics for the bees that are still alive. This can be prevented by making sure that you get rid of all honey from the hive before the cold sets in. As this will lower the chance of developing dysentery, you can also give the honey syrup to the bees late in the autumn.

Small Hive Beetle

Despite being small, the Small Hive Beetle can cause serious damage. It can not only destroy the bees, but also cause damage to the honey, comb and pollen. Infestation may be so severe that bees are forced to leave their hive.

If this problem is not addressed, honey can be destroyed. Small Hive Beetle larvae tunnel through honey to feed on it. They will also leave behind feces which can cause honey to turn brown and ferment its taste. You will notice an orange hue to honey when this happens. You will also notice the same smell that rotting oranges.

Treatment & Prevention

To treat a problem caused by the Small Hive Beetle, it is important to get to the root of the problem. They live their entire life cycle in the ground surrounding the hive. It is important to prevent ants from entering the hive. This can be done by placing diatomaceous earth around the hive to stop them from becoming a problem. They can become dangerous to the hive if they become too dry.

If you find evidence of the Small Hive Beetle, there are many pesticides that can successfully be used. The pesticide is applied to cardboard to allow the beetles to enter the cardboard, but the honey bees cannot as they don't want to reach the pesticide. If you are looking for a non-chemical way to get rid of the beetles, bottom board traps made from cooking oil can also work.

Wax Moths

Wax moths are most common in honey bees. Although they don't directly harm bees, they do feed on honeycomb wax. Bees' survival and the preservation of their honeycomb is at risk. The honey stored in the honey can be contaminated by wax moths. They can also destroy bee larvae.

Treatment & Prevention

Wax moths can be easily controlled if you are able to quickly take care of them. You can kill the larvae and eggs by freezing, so make sure to store your hives properly in winter. Wax moths can be a problem if the hives are placed in heated storage.

The bees will remove moth larvae and webs as long as they are healthy and strong. You can use various tools to remove cell accumulations so honey bees don't become ill. You can also use moth crystals and a urinal disk to get rid of them. Mothballs are not recommended as they can kill the bees and contaminate honey from wax accumulation.

Chapter 16: Beekeeping Equipment

You should have basic tools to help you collect honey and manage your beehives. Bees love dim colors. Therefore, the best cloth is one that is light in color. Perfumes, colognes, and scented lotions can also attract bees. Avoid them.

Standard beekeeping needs special equipment and tools that are specifically designed to protect against the sting. These tools include:

Smoker

This is a must-have tool to keep bees. This is a container made of metal with attached bellows. It is used to inhale smoke onto the hives.

Beekeeping Brush.

Beekeepers are advised to use a very soft hairbrush to remove bees from their hives to prevent cell damage. The most common is horsehair.

Scraper/Hive tool.

These tools are essential for removing hive parts, manipulating frames and scraping wax. A popular tool to keep bees is the hive tool or scraper.

Knife or wax cutter

You can use either an electric or a normal wax cutter. Before you put in the extractor, you will need to remove the cells that are capped to make honey. While an electric knife is the most active, you could also use an uncapping roll, fork, or sharp knife.

Heating pot for wax.

To make a wax sheet you need to melt the wax. You can make the wax sheet at home or buy it from beekeeping shops.

Make wax sheets

This tool can be used to create a foundation for a beehive. It will assist bees in building their nest as quickly and easily as possible. Their nest building performance is enhanced by wax foundation.

Comb puller.

Propolis is often used to fasten bee comb. It is very difficult to remove with your naked hand. The task of removing comb is made easier by using a comb puller, even for sticky combs.

Beehive Equipment

You can order hives online, or buy them from local beekeepers. Hives are a vital tool for beekeepers. To build bee hives, you will need all the necessary components and tools, such as foundation pins, nails, boxes, and starter kits. They are easy to assemble and can be purchased at any time. Your hives will have various components, such as top and bottoms bars and net trap.

You will need to inspect your hives on a regular basis. Before you start beekeeping, it is advisable to practice assembling and disassembling the honeybee hives.

Beehive Frames & Feeder

You must now have your bees and your beehives installed. If nectar isn't available at the time you start, install a honeybee feeder to provide food for your bees. One feeder should be placed at the entrance to the hive, and one on the edge of direct brood chamber. A 3rd pail feeder will be on the inner cover hole. The 4th baggy feeder will be on the planks beside the top bars. You can place the 5th feeder in the deeper hive.

Install a feeder and beehives. The feeder is an important part of the hive for the first few weeks. Keep in mind that not all hives have the same options for feeder installation. One feeder is sufficient for one hive structure.

These parts make up a beehive:

Frames superbox. You should place 10 frames superbox on top of the 10 frames brood box.

Brood chamber/box This is the base and bottom of beehives. This can have eight to ten frames.

Super box. Super box. The super box frames should be equal or less than the brood chamber in number.

Top cover. To protect against rain and wind, the top cover is placed on the hive after it has been installed. To keep the body together, the bees will use propolis from trees to fasten it. If there is a slight gap or hole in the body or cover, they will enclose it with wax.

Separator between super box and brood. This separator is used to separate the queen from the eggs that were laid in the superbox during harvesting. The super box is used for honey harvest. It can be difficult to remove if eggs are being laid by the queen and it may become damaged.

Eight to ten frames can be attached with a wax sheet (prepared manually or by bees). Two types of boxes can be used for eight to ten frames. Wax sheet is a synthetic beehive that beekeepers make with wax to help them complete their natural hive.

To manually feed in the off-season, use a feeder pot. This is usually used in the off-season, when there is a lack of nectar and flowers. Beekeepers will keep sugar syrup in a feeder pot every day as an alternative feed.

Beekeeping Suit & Veil

A beekeeping suit is a cloth that covers your trunk and limbs. It is usually made from light-colored materials. A beekeeping veil and suit are required to protect your skin and body from the bee stings. These protective devices are essential for beginners in beekeeping. Also, gloves and shoes should be worn. Remember that bees can climb up your clothes if they get into your fabric. You may have to cover any cracks with string or masking tape. While you can eventually wear regular cloth instead of a beekeeping suit, as a beginner it is best to use a mask to protect your face, neck, and eyes.

These are the beekeeping suits:

The beekeeper's veil: This will protect your head, shoulders, neck, neck and face from the bee's sting.

The full shirt and pants will protect your entire body from the bee sting. To ensure safety, make sure there is no space between your body and the cloth.

Safety shoes: These safety shoes will protect your legs against the sting of bees and make it easier to handle beehives. Your pant should always be in the safety shoes.

Gloves for Hands: To ensure the safety of your hands while working on the beehive, gloves are essential.

10. 10.

Honey extractor or manual harvester: You can extract honey manually if you have several beehives. This is a good option for beekeepers who are part-time.

Automated Honey Extractor: This tool is ideal for professional beekeepers and large bee farms. When you have thousands or more beehives, it is nearly impossible to extract honey manually.

Sugar syrup making pot: When there are not enough nectar-producing flowers, sugar syrup is an option. This will keep your bees healthy and happy.

To clean syrup or honey, a sieve or filter is used. A lot of honey and wax can be mixed together during honey extraction. To separate them, you need to filter the honey thoroughly.

Funnel/cone. This can be used to quickly and easily fill a jar or bucket with honey.

11. Honey (Manual Processing Equipment)

Heater: For manual processing, a heater with a control switch is crucial

Wooden separators and pots that can be used between two pots (inner or outer). These wooden separators can be used to seperate two pots to protect them from direct heat, which could cause damage to several honey ingredients.

Thermometer or temperature indicator: This thermometer is used to regulate heat and to prevent excessive heat.

Moisture indicator. This indicator is used to monitor honey thickness and moisture, or to maintain a moderate amount of honey moisture. The moisture range for honey is eighteen to twenty two percent.

Manual processing requires the use of the 4 items mentioned above.

12.   Honey   Processing   Machine (Automatic)

Because it is very expensive, this is not recommended   for   commercial beekeepers. A small power blanket is an option, and it's very affordable.

Chapter 17: Honey Processing

Supers could be taken out and processed in a perfect world. Honey can also be robbing by bugs or mice, wax moth damage, and fermentation.

Supers can be stored in a garage or an outdoor workshop, as long as it is dry and free from excess heat.

Honey stored in supers can still be at risk from ants, bees, and wasps.

Honey's main ingredient is its water content.

There are some things you need to do in a honey room.

Honey's thickness changes with temperature. The thicker honey, the more it will melt, the harder it is to extract from the extractor.

When the room gets cooler, honey and wax will eventually reach every corner.

There are many options for beekeepers who just started out.

Use honey from the comb.

Section honey.

Honey extracted.

Cut comb honey can be removed from the frame and placed in 8 oz. Each package contains 12 oz. It comes in pieces.

A common use for cut comb containers is to hold a comb measuring 40mm.

It is thick.

Examine the frame before cutting to decide which side of the comb has the better appearance.Lay the frame on a clean tray, and the whole comb cut out of the frame with a sharp knife.Only the best parts of the comb can be used.The hollow parts at the edge should not be used and uncapped cells kept to a minimum.A sharp kitchen knife, a cheese wire, or a stainless steel comb cutter can be used to cut the combs.All portions of cut comb should stand on a grid to let the honey drain from the outside cut cells.A piece of comb honey swimming in its container in liquid honey is poor presentation.Because heather honey is a gel it can be packaged straight away.The best storage for comb honey is in a deep freeze, in special plastic boxes, where comb will keep indefinitely.Freezing packaged comb honey will also kill any wax moth eggs and larvae.Comb honey stored in any other fashion must be examined regularly for signs of deterioration.Another development of comb honey is chunk honey.Chunk honey is a piece of cut comb is put in a jar and surrounded with a clear runny honey, producing what is am attractive presentation.

Extracting wax cappings is a very valuable side product.

Section honey is a skillful bee craftman.

You can extract honey from honey without a centrifugal extractor. All you need is basic kitchen tools to deal with the supers.

The comb, cells, comb and honey must be removed from the frame using this method.

There are some disadvantages.

Heat can remove some compounds that give honey its unique flavor.

The wax will begin to soften, making it more difficult to uncapping, as the cell walls are dragged along with the knife. At 450C, combs will start to soften and then collapse. At 630C, wax will melt.

Each frame is removed from the super using one lug. This varies depending on how the uncappings were done.

Uncapping the honey into a bucket, basin, or uncapping tray is the easiest way to do this. Then, gravity strain the honey with a strainer, sieve, or strainer. A filter bag that can hold 70 lb. Plastic tanks are often used. The honey in the wax caps can be used to make honey wine or be fed back to bees.

The honey and wax can be separated using a heated tray.

Other processes are possible to separate honey and wax, but these require more sophisticated equipment.

Equipment for Honey Processing

The principle of a centrifugal extractor is the same as a centrifuge.

Let's start with tangential.

This machine has a short handling time. However, it is capable of extracting honey from heather.

The frames are placed between rings and are arranged in the same way as the spokes of a radial wheel. However, the radial machines can handle more frames in the reverse direction to extract more honey from the cells.

Tin-plated steel was the traditional material used to construct the machines. However, once it starts rusting, there are very few options. Stainless steel is much more durable than plastic.

A manual extractor can be used if you only have two or three bees to extract honey. An electric extractor can be used for larger quantities.

Chapter 18: Getting the Bees

If you're a beginner beekeeper, the first thing you should do is to put your bees into their new hives. Enjoy the process. Contrary to common belief, bees are cooperative and docile creatures. They will not sting if they are disturbed. To be a beekeeper, the first step is to make sure your bees are comfortable in their new environment. Make sure you read the instructions multiple times until you feel confident with them. Also, make sure you do at least one dry run before you actually start.

A reputable breeder is the best place to purchase your bees. It is important to consider the environment where you will be raising bees. This includes their viability, adaptability, and reproducing ability in that particular climate. You want gentle bees.

You should ideally hive your bees by the afternoon of the day you picked them up or the next morning. Expert beekeepers agree that it is best to transfer bees to a new home in the evening or late afternoon. It's the best time for them to go back to the hive to settle down for the night. You may need to wait until it is cooler or windier.

Choose a day that is clear and mild with very little wind. You should wait until it stops raining or cold to avoid your insects becoming agitated.

These are the basic steps you should follow. (Blackiston, 2015)

30 minutes prior to hiving, spray your honeybees with non-medicated sugar syrup. But don't drown them in syrup.

Use your hive tool to pry the wooden cover from the package. Keep the wood cover safe by removing the staples and nails.

Place the package on its bottom and jar it down so your bees can reach the bottom. They will not feel any harm! Take out the syrup can from the box and remove the queen cage.

Take a look at the queen cage. Verify that she is OK. Rarely, she might have died during transit. If this is the case, proceed with the installation as though everything was fine. Call your supplier to request a replacement queen. It should not cost you anything. While you wait for the replacement queen, your colony will be okay.

Slowly slide the metal disc from the queen cage to one side. You can see the white candy inside the hole if you remove the cork from one end. Remove the entire disc if the candy is still present. You can fill the gap with a piece of marshmallow if the candy is not present.

Make a hanging bracket to hang the queen's cage using two small frames nails that have been bent at the right angles.

Spray your bees once more, then jar the container so that the bees sink to the bottom.

Remove five frames from the hive and place them in a nearby location.

At this stage, you are only using the deeper hive bodies for your bees. Hang the queen cage, candy side up, between the middle-most frame (facing toward the center) and the next frame. The cage screen should face towards the middle of the hive.

Conclusion

We are grateful that you downloaded this book.

I hope you found this book helpful in understanding the many advantages and disadvantages of beekeeping. This book is primarily for people who are interested setting up their own beekeeping apiaries. Everybody has to start somewhere. This book will give you an overview of the many nuances involved in beekeeping. While you won't make a profit the first year, you can start to see results as you grow your business. This book will help you understand the importance and importance of proper design and how to take care of your hives by only meeting her needs. The book includes a good recipe for good food for bees and explains the importance of feeding the colony.

We also talked about the design of the beehives. And we even discussed briefly the best bee species for beginners. These steps will help you establish a healthy and happy bee colony in your yard. We also discussed the many ways that you can harvest the honey.

We also discuss safety in beekeeping and provide basic information about the most important safety equipment. Finally, we will talk about the cost of setting up beehives. It is expensive to start, and you will only get 75 pounds of honey in your first year. However, this won't be enough to make it profitable. Beekeeping is just like any other farming method. To reap substantial profits, you must be patient.

Next, you will need to try beekeeping for yourself. Before you dive into beekeeping, you should learn about the seasons in your area. Also, know when the best time is to start your business. You should also do enough research to determine the best bee species to breed in your area. There are many species of bees that can be found in different climates. Therefore, it makes sense to find the best bee species for your climate. Talk to an experienced beekeeper regarding the dangers involved in beekeeping. Experience is the best teacher. You can observe his daily work and stay with him on the bee farm for a few days to get a better understanding of beekeeping.

www.ingramcontent.com/pod-product-compliance
Lightning Source LLC
Chambersburg PA
CBHW071212210326
41597CB00016B/1787